R. M. Bulgin

FALCON PROSE CLASSICS
General Editor : Leonard Russell

★

ROBERT SMITH SURTEES

Robert Smith
SURTEES

SCENES AND CHARACTERS

edited and introduced by
CYRIL RAY

London
THE FALCON PRESS

*First published in 1948
by the Falcon Press (London) Limited
7 Crown Passage, Pall Mall, London S.W.1
Printed in Great Britain
by Hazell, Watson & Viney, Ltd.
London and Aylesbury
All rights reserved*

CONTENTS

INTRODUCTION	*page* 7
BIOGRAPHICAL NOTE	13
A BRIEF BIBLIOGRAPHY	16
INTRODUCING MR. JORROCKS	17
THE COCKNEY SPORTSMAN	22
THE MAN WHO CAME TO DINNER	26
CLASSIC DISCOVERY	35
ENTER MR. SPONGE	37
THE SPONGE CIGAR AND BETTING ROOMS	42
EARLY-VICTORIAN EXQUISITE, MALE	46
EARLY-VICTORIAN EXQUISITE, FEMALE	49
ENGLISH MISS, C. 1845	51
BEDROOM SCENE	53
PROFESSIONAL POLITICIAN	57
A LITTLE POLITICAL ACCOMMODATION	68
GRASS-LAND AND ARABLE	78
ENTER THE RAILWAY TRAIN, AND THE LONDON CLUB	80
THE WOTHERSPOONS AWAIT THEIR GUESTS	89
SOCIAL CALL	91

INTRODUCTION

'DICKENS and horsedung' is the description given to Surtees's novels by a Kipling character, though the same man, we are told, became fascinated by the 'heavy-eating, hard-drinking hell of horse-copers, swindlers, match-making mothers, and economically dependent virgins selling themselves blushingly for cash and lands'. A foul world, the Kipling character calls it, until he recognizes it as a real one, and learns to identify his neighbours, one by one, 'out of the natural-history books by Mr. Surtees'.

There is some sound criticism in that not very pleasant Kipling story, though 'Dickens and horsedung' is about as fair and comprehensive a summing-up of Surtees as 'Surtees and saccharin' would be of Dickens. There is no profit, in any case, in these literary comparisons, though if the modest Surtees must be balanced against the giants of his period it is worth remembering that Thackeray envied him his power of characterization, and that William Morris (of all people, to praise the novels of a sporting squire,) ranked him with Dickens as 'a master of life'.

What Kipling had left out of account, of course, is the salty, boisterous humour of the Surtees novels. Surtees's world of shady horse-copers and shadier horses may be a foul world but it is a funny one. Soapey Sponge is on the make, and Lucy Glitters may be no better than she should be (though the one literary convention of the time that Surtees almost respects is the sex tabu), but they are comic figures and not exercises in realism. And one, at least, of the Surtees characters is a comic figure in the grand manner, with a niche beside those of Sam Weller, Mr. Polly, and the Good Soldier Schweik.

It is because of John Jorrocks (and Pigg, for the two are as inseparable as Quixote and Sancho Panza—though you would be hard put to it to decide which was the Sancho and which the Quixote)—it is because of Jorrocks that *Handley Cross* is the best loved and most quoted of Surtees's novels. It is not the best novel in its own right; *Mr. Sponge's Sporting Tour* is a more skilful, a more professional, if not a more satisfactory piece of work. Surtees's contemporaries were of the same opinion. Jorrocks was created in 1831, in sketches in the *New Sporting Magazine*, when Surtees was twenty-six and a sporting journalist. (He had qualified in London as a solicitor and was soon to revert to type in County Durham, by virtue of inheritance, as a country gentleman and master of foxhounds.) Jorrocks's earlier adventures, published in book form as *Jorrocks's Jaunts and Jollities*, cut no ice, and hardly deserved to. But neither did *Handley Cross* itself (published in three volumes in 1843), though Jorrocks, the Cockney sportsman, gave someone in the meantime the idea that blossomed into *The Pickwick Papers*. It was ten years before Surtees wrote a really successful novel: *Mr. Sponge's Sporting Tour*. *Handley Cross* was of no importance until it was republished (in 1854) with the success of *Sponge* and the new, richly comic illustrations by Leech to send it off. It had taken Jorrocks nearly a quarter of a century to come into his own.

It would be hard now to understand why, if it were not that Jorrocks is still known of, rather than known, by many who have otherwise read lovingly in the period. The admirable bibliography, for instance, in Trevelyan's *Nineteenth Century* quotes Arthur Young, Cobbett, Jane Austen, Peacock, Dickens, Borrow, Mrs. Gaskell, Disraeli, Trollope, Meredith—even Colonel Hawker's *Instructions to Young Sportsmen*—on the century's social life alone and says never a word of Surtees, though in the richness of his detail he is the best guide of all to the clothes, transport, food, furniture, and fun enjoyed in England between the first and second Reform Bills. Many have fought shy of Jorrocks—as I did, for too long—because they detested blood sports. It would be as sensible to avoid Sherlock Holmes or Lord Peter Wimsey because of an objection to capital punishment.

In any case, Surtees is not always preoccupied with hunting. Mr. Jorrocks gives up the Handley Cross mastership to become the hero of *Hillingdon Hall*, and the theme of that book is agricultural improvement and the anti-Corn-Law movement; *Ask Mamma* is as much about the marriage market and about the changing face of England—the steam locomotive was driving the coach from the landscape—as about the hunting-field; *Plain or Ringlets?* as much about the seaside promenade, the new railways (again), the even newer crinoline and—the title is a give-away—the newest fashions in what we now call hair-does.

And even in the fox-hunting novels the hunting-field is only a background for Surtees's cynical sense of fun, for his fertile invention of character, his acute observation of dress and manner and idiom. His highly-coloured, richly individual men and women are never overshadowed by horse or by hound—eccentrics though some of the four-legged characters are in their own right. If it was in creating Jorrocks that Surtees's agreeable talents blossomed into something approaching genius, the fact is recognizable even by those who have never seen a hound or been across a horse.

A nodding acquaintance with horses, though, does bring the fun nearer home. For Jorrocks, devoted though he is to hunting—'all time is lost wot is not spent in 'unting—it is like the hair we breathe—if we have it not we die—it's the sport of kings, the image of war without its guilt, and only five-and-twenty per cent. of its danger'—devoted though he is, he is no horseman. That is Pigg's value in the book, to show off by his hard riding the apprehensive clumsiness of his master. Throughout the book Jorrocks is having horse trouble, coming off over the head of a standing horse when the great sporting journalist is there to immortalize his hounds, or roaring "Old up! 'oss', holding reins in one hand and saddle with the other, as Arterxerxes crashes sideways through a fence that Pigg had taken like a bird.

It is not that Jorrocks was a stranger to the hunting field when he took over the hounds at Handley Cross. Had he not served his apprenticeship, like the other Cockney sportsmen of the period, with the Surrey Hunt? And, when the railways came, living as he did in Great Coram Street, did he not find that the two best cover hacks in the world, the Great Northern and Euston stations, were at his doorstep to take him eventually into the foxhunter's heaven of Leicestershire? It was there, in the 'cut 'em down countries', that Jorrocks could be seen hobnobbing with the 'hupper crusts . . . until he got fairly launched among them—when he would out with his order-book and do no end of business in tea.'

For Jorrocks is a grocer, with a business in the city and a house in Bloomsbury. And it was this, perhaps, that made *Handley Cross* hang fire. One biographer, at any rate, Frederick Watson, has pointed out that when the book first appeared the social predominance of foxhunting was at its height, and that the M.F.H. of the day, modelling himself, no doubt, on Lord Althorp of the Pytchley, could not take kindly to a Cockney grocer as a brother master. It would have been different had Surtees made Jorrocks simply a figure of fun for gentlemen in the shires to laugh at. Jorrocks is a figure of fun, sure enough, but the gentlemen, the cut 'em down captains and the Sir Archey Depecardes, come in for rougher handling. Surtees makes fun of Jorrocks, but it is affectionate fun; Surtees is no snob.

Nor is Jorrocks, or he would not have flourished his order-book in the Midlands. Hunting to him is not a short cut to Society; it is a proper exercise and hobby—a proper passion in his case—for a well-to-do man of business. He has no social pretensions: he is a Post Hoffice Directory, not a Peerage, man, and explains that 'there are but two sorts o' folks i' the world, Peerage folks, wot think it's right and proper to do their tailors, and Post Hoffice Directory folks wot think it's the greatest sin under the sun not to pay twenty shillins i' the pund—greatest sin under the sun 'cept kissin' and then tellin'.' He takes on the mastership at Handley Cross, though a mastership he admits to be a werry high step in the ladder of his hambition, only after cautious inquiry into the cost of hunting the country and the financial standing of the subscribers.

He is a character in the round and all of a piece. Commonsense never fails him—commonsense and a superb natural judgment. Make him a present of a horse and he will give you his opinion of a present wot eats. He is offered mince at dinner and his answer is, 'No, I likes to chew my own meat.' His own tastes have the simple magnificence that one expects of a merchant prince: his 'best dinner that ever was cooked' was turtle soup and turbot, a haunch of doe venison and Stilton. And where he dined he slept, and where he slept he breakfasted. Throughout the book Jorrocks works his way through mountains of food, seas of port, and oceans of brandy and water. Drinking must not interfere with sport (and here Jorrocks and Pigg join issue, for Pigg is known to complain that Jorrocks has no business out a-huntin' on a drinkin' day), although Jorrocks can exclaim, shivering before a jump, 'Fancy a great sixteen 'and 'oss lyin' on one like a blanket, or sittin' with his monstrous hemispheres on one's chest; sendin' one's werry soul out o' one's nostrils. Dreadful thought! Vere's the brandy?' But when there is no hunting then is the time for conviviality, on every kind of drink. It is on brandy and water that the pair so fuddle themselves that Pigg, sent to look out of the window to see what sort of night it is, makes the classic discovery that it is hellish dark and smells of cheese. It is the harder-headed Jorrocks that points out, 'Vy, man, you've got your nob i' the cupboard!'

Jorrocks is hard-headed in every sense. You never have any doubts about his ability, as you have of Pickwick's, to have earned his reputed fortune in the City; he will never tip young Binjimin with a pot of marmalade out of stock if there is half a pot handy. It is Binjimin, by the way, who is the chopping-block for much of Jorrocks's fine impromptu eloquence. 'Come on, ye miserable useless son of a lily-livered besom-maker,' he bellows at him in the hunting field. 'Rot ye, I'll bind ye 'prentice to a salmon pickler.' But many gentler folk come off no better. He roars objurgations to 'You 'airdresser on the chestnut 'oss,' who turns in fury with, 'Hairdresser! I'm an officer in the ninety-first regiment!' 'Then, you hossifer in the ninety-fust regiment, wot looks like a 'airdresser, 'old 'ard!'

Handley Cross has obvious faults. It is a sprawling, plotless novel, too long by half and quite out of balance: Jorrocks himself does not appear until the seventh chapter, Pigg not until the twentieth. The names of the minor characters are far too farcically apposite—as bad as anything in Dickens and far worse than Trollope's Quiverful or Thackeray's almost forgive-able Dobbin (Miserrimus Doleful is too much for the strongest stomach). The maddening convention is observed of having

Jorrocks replace aspirates with apostrophes in written letters. In spite of this, and even without Jorrocks, the gusto of the writing is disarming, though without Jorrocks it would be no more than a mine of material about English social life in a strong-flavoured, high-coloured age. Jorrocks makes it a sort of blundering masterpiece.

The other novels are less than that, yet well worth the reading. Surtees reached London in the eighteen-twenties, died in 1864; his productive period lay between the Regency, 'full-blooded yet intellectual, aristocratic and at the same time slightly vulgar,' (as Mr. Osbert Lancaster observes of the furniture of the period) and the more hypocritical, more conventional, yet no less prodigal early Victorian age. He stands between the golden era of the stage-coach and the railways virtually as we know them, between Pierce Egan's *Life in London* and the London, almost, of Henry James. There has never been in England such a forty years of change, such a period of gross eating, fancy dressing, and personal eccentricity. And Surtees packs them all in: guzzlers like Mr. Jovey Jessop, in *Plain or Ringlets?* with his English cook for the breakfast beefsteak and his French cook for the fricandeau for dinner; bobby-dazzlers like Laura Guineafowle, in *Young Tom Hall*, with her beautiful light blue silk evening gown, its trimming *en tablier*, its narrow silk flounces embroidered in chain stitch, its 'fly-away sleeves for sweeping things off tables and draggling into teacups and soup-plates'; rum 'uns like breathless little Jogglebury Crowdey, in *Mr. Sponge*, providing for his family by carving half the hedgerows in the county into 'curious-handled walking-sticks'.

It is a knowing, amused, disillusioned eye that the dry north-country squire cocks at them. Surtees never expected too much of his fellows, and was never disappointed. There is none of Thackeray's moralizing in his novels, none of Dickens's namby-pamby heroes or milk-and-water heroines. Let us be frank: Surtees is so much smaller a man than they, that if there were he would be unreadable.

As it is, though, his novels know their place, as it were; they are good fun because they serve no social purpose, and their characters are amusing because nobody asks you to admire them. They are as much above Samuel Warren's *Ten Thousand a Year* as they fall short of *Pickwick*. They are as English as bitter beer, but as dry as Tio Pepe. And no English novelist that I know of has provided such a meticulously detailed background—town, country, railway-station, cigar-divan, dining-room; furniture, clothes, plate, china, guttering candles;—for such a diverting crew of scamps, fools, husband-hunting heroines, and malicious old trout. Surtees was the last novelist for half a century to make his hero, as in *Mr. Sponge*, 'tolerably sharp' and his heroine 'tolerably virtuous'. There is a clear eye here, looking at people as they are; that alone should commend Surtees to our own disillusioned age.

BIOGRAPHICAL NOTE

Robert Smith Surtees was born in 1803, the second son of Anthony Surtees, of Hamsterley Hall, Co. Durham, and a connection of Robert Surtees, the antiquary friend of Sir Walter Scott.

Like the sons of many other country gentlemen, before Arnold and the railways made the modern public-school system, Surtees was educated at the local grammar school, and then sought a younger son's profession in a solicitor's office in London.

Surtees was already an admitted solicitor and, more important, a sporting journalist when in 1831 the death of his elder brother, unmarried, made him the heir to a considerable family property and altered the course of his life. The same year saw him join Rudolph Ackermann, the younger, in founding the *New Sporting Magazine*, in the pages of which he created John Jorrocks, the hunting grocer, as the hero of a series of sporting sketches. He had just published *The Horseman's Manual*, in which he used his legal knowledge and his experience of horses and horse-copers to construct a cautionary guide for buyers.

In 1838, on the death of his father, Surtees succeeded to Hamsterley Hall and to the almost inevitable career, for a country gentleman, of justice of the peace, Master of Foxhounds, militia major, and High Sheriff of the county. But with this difference. His sketches from the *New Sporting Magazine* collected and published as *Jorrocks's Jaunts and Jollities*, were successful enough, not only to give Chapman and Seymour the idea for a similar series about Cockney sportsmen that led to *Pickwick*, but also to lead Lockhart and "Nimrod" (Apperley, the hunting correspondent) to persuade Surtees to embark upon *Handley Cross*.

So Surtees continued author, though with a characteristic clinging to his amateur status; *The Horseman's Manual* was the only work to which he ever put his name. But the novels came out steadily for the next quarter of a century, first as serials and then in book form; and Surtees found time, too, to join in founding *The Field*.

It was a quiet, uneventful life that Surtees led in County Durham, whence he watched the innovations of an inventive age with a conservative, but a not too prejudiced, eye. He was as go-ahead about railways and agricultural improvement as he was cynical about the abolition of slavery and the Corn Laws; he could be as scornful about the British Army as any pacifist radical—and then twice as scornful about the Radicals.

Tall, lean, and a good horseman, Surtees was as unspectacular in the hunting field as he was retiring as an author. "Without ever riding for effect, he usually saw a good deal of what hounds were doing." Much the same can be said of Surtees as a novelist of contemporary country life. He was well enough known to his contemporaries—Thackeray, Harrison Ainsworth, and the rest —known and liked, but they have told us as little about him as he told them about himself.

Surtees died in 1864. He had married, in 1841, a neighbouring heiress; his only son died young, but one of his two daughters married, in 1885, John Prendergast Vereker, heir to the viscountcy of Gort, and the present Lord Gort is his grandson.

A BRIEF BIBLIOGRAPHY

[I have done as Mr. Michael Sadleir did with Trollope: recommended books are marked with an asterisk—a few, Baedeker fashion, with two.]

THE MAJOR WORKS OF SURTEES

(Dates are of first publication in book form.)
Jorrocks's Jaunts and Jollities. 1838.
***Handley Cross.* 1843.
Hillingdon Hall. 1845.
The Analysis of the Hunting Field. 1846.
Hawbuck Grange. 1847.
***Mr. Sponge's Sporting Tour.* 1853.
**Ask Mamma.* 1858.
**Plain or Ringlets?* 1860.
**Mr. Facey Romford's Hounds.* 1865.
Young Tom Hall. An unfinished novel, rescued by Mr. E. D. Cuming from the *New Monthly Magazine*, in which it appeared as a serial, and published in 1926.

BOOKS ON SURTEES AND HIS WORKS

**Robert Smith Surtees*, by himself and E. D. Cuming, 1924.
***Jorrocks's England*, by Anthony Steel. 1932.
**Robert Smith Surtees*, a critical study, by Frederick Watson. 1933.

INTRODUCING MR. JORROCKS

MR. JORROCKS was a great city grocer of the old school, one who was neither ashamed of his trade, nor of carrying it on in a dingy warehouse that would shock the managers of the fine mahogany-countered, gilt-canistered, puffing, poet-keeping establishments of modern times. He had been in business long enough to remember each succeeding Lord Mayor before he was anybody—'reg'lar little tuppences in fact', as he used to say. Not that Mr. Jorrocks decried the dignity of civic honour, but his ambition took a different turn. He was for the field, not the forum.

As a merchant he stood high—country traders took his teas without tasting, and his bills were as good as bank-notes. Though an unlettered man, he had great powers of thought and expression in his peculiar way. He was 'highly respectable', as they say on 'Change—that is to say, he was very rich, the result of prudence and economy—not that he was stingy, but his income outstripped his expenses, and money, like snow, rolls up amazingly fast.

A natural-born sportsman, his lot being cast behind a counter instead of in the country, is one of those frolics of fortune that there is no accounting for. To remedy the error of the blind goddess, Mr. Jorrocks had taken to hunting as soon as he could keep a horse, and although his exploits were long confined to the suburban county of Surrey, he should rather be 'credited' for keenness in following the sport in so unpropitious a region, than 'debited' as a Cockney and laughed at for his pains. But here the old adage of 'where ignorance is bliss', etc., came to his aid, for before he had seen any better country than Surrey, he was impressed with the conviction that it was the 'werry best', and their hounds the finest in England.

'Doesn't the best of everything come to London?' he would ask, 'and doesn't it follow as nattaral consequence, that the best 'unting is to be had from it?'

Moreover, Mr. Jorrocks looked upon Surrey as the peculiar province of Cockneys—we beg pardon—Londoners. His earliest recollections carried him back to the days of Alderman Harley, and though his participation in the sport consisted in reading the meets in a bootmaker's window in the Borough, he could tell of all the succeeding masters, and criticize the establishments of Clayton, Snow, Maberly, and the renowned Daniel Haigh.

It was during the career of the latter great sportsman that Mr. Jorrocks shone a brilliant meteor in the Surrey hunt—he was no rider, but with an almost intuitive knowledge of the run of a fox, would take off his hat to him several times in the course of a run. No Saturday seemed perfect unless Mr. Jorrocks was there; and his great chestnut horse, with his master's coat-laps flying out beyond his tail, will long be remembered on the outline of the Surrey hills. These are recollections that many will enjoy, nor will their interest be diminished as time throws them back in the distance. Many bold sportsmen, now laid on the shelf, and many a bold one still going, will glow with animation at the thoughts of the sport they shared in with him.

Of the start before day-break—the cries of the cads—the mirth of the lads—the breakfasts at Croydon—the dear 'Derby Arms'—the cheery Charley Morton; then the ride to the meet—the jovial greeting—the glorious find, and the exhilarating scrambles up and down the Surrey hills.—Then if they killed!—O, joy! unutterable joy! How they holloaed! How they hooped! How they lugged out their half-crowns for Tom Hill, and returned to town flushed with victory and 'eau-de-vie'.

But we wander—'

When the gates of the world were opened by railways, our friend's active mind saw that business might be combined with pleasure, and as first one line opened and then another, he shot down into the different countries—bags and all—Beckford in one pocket—order book in the other—hunting one day and selling teas another. Nay, he sometimes did both together, and they tell a story of him in Wiltshire, holloaing out to a man who had taken a fence to get rid of him, 'Did you say *two* chests o' black and *one* o' green?'

Then when the Great Northern opened he took a turn down to Peterborough, and emboldened by what he saw with Lord Fitzwilliam, he at length ventured right into the heaven of heavens—the grass—or what he calls the 'cut 'em down' countries.[1] What a commotion he caused! Which is Jorrocks? Show me Jorrocks! Is that old Jorrocks? And men would ride to and fro eyeing him as if he were a wild beast. Gradually the bolder ventured a word at him—observed it was a fine day—asked him how he liked their country, or their hounds. Next, perhaps, the M.F.H. would give him a friendly lift—say 'Good morning, Mr. Jorrocks'—then some of what Jorrocks calls the 'hupper crusts' of the hunt would begin talking to him, until he got fairly launched among them—when he would out with his order-book and do no end of business in tea. None but Jorrocks & Co.'s tea goes down in the Midland counties. Great, however, as he is in the country, he is equally famous in London, where his 'Readings in Beckford' and sporting lectures in Oxenden Street procured him the attentions of the police.

Mr. Jorrocks had now passed the grand climacteric, and balancing his age with less accuracy than he balanced his books, called himself somewhere between fifty and sixty. He wouldn't own to three pund, as he called sixty, at any price. Neither could he ever be persuaded to get into the scales to see whether he was nearer eighteen 'stun' or twenty. He was always "ticlarly engaged' just at the time, either goin' to wet samples of tea with his traveller, or with someone to look at 'an 'oss', or, if hard pressed, to take Mrs. J. out in the chay. 'He didn't ride stipple chases', he would say, 'and wot matter did it make 'ow much he weighed? It was altogether 'twixt him and his 'oss, and weighin' wouldn't make him any lighter'. In person he was a stiff, square-built, middle-sized man, with a thick neck and a large round head. A woolly broad-brimmed lowish-crowned hat sat with a jaunty sidelong sort of air upon a bushy nut-brown wig, worn for comfort and not deception. Indeed his grey whiskers would have acted as a contradiction if he had, but deception formed no part of Mr. Jorrocks's character. He had a fine open countenance, and though his turn-up nose, little

[1] "Cut 'em down and hang 'em up to dry!" Leicestershire phrase.

grey eyes, and rather twisted mouth were not handsome, still there was a combination of fun and good humour in his looks that pleased at first sight, and made one forget all the rest. His dress was generally the same—a puddingy white neckcloth tied in a knot, capacious shirt frill (shirt made without collars), a single-breasted, high-collared, buff waistcoat with covered buttons, a blue coat with metal ones, dark blue stocking-net pantaloons, and hessian boots with large tassels, displaying the liberal dimensions of his full, well-turned limbs. The coat pockets were outside, and the back buttons far apart.

His business place was in St. Botolph's Lane, in the City, but his residence was in Great Coram Street. This is rather a curious locality—city people considering it west, while those in the west consider it east. The fact is, that Great Coram Street is somewhere about the centre of London, near the London University, and not a great way from the Euston station of the Birmingham railway. Jorrocks says it is close to the two best cover hacks in the world, the great Northern and Euston stations. Approaching it from the east, which seems the proper way of advancing to a city man's residence, you pass the Foundling Hospital in Guildford Street, cross Brunswick Square, and turning short to the left you find yourself in 'Great Coram Street'. Neat unassuming houses form the sides, and the west end is graced with a building that acts the double part of a reading-room and swimming-bath; 'literature and lavement' is over the door.

In this region the dazzling glare of civic pomp and courtly state are equally unknown. Fifteen-year-old footboys in cotton velveteens and variously fitting coats being the objects of ambition, while the rattling of pewter pots about four o'clock denotes the usual dinner hour. It is a nice quiet street, highly popular with Punch and other public characters. A smart confectioner's in the neighbourhood leads one to suppose that it is a favourite locality for citizens.

Mrs. Jorrocks, who, her husband said, had a cross of blood in her, her sire being a gent, her dam a lady's maid, was a commonish sort of woman, with great pretension, and smattering of gentility. She had been reckoned a beauty at Tooting but had outlived all, save the recollection of it. She was a dumpy figure, very fond of fine bonnets, and dressed so differently, that Mr. Jorrocks himself sometimes did not know her. Her main characteristics were a red snub nose, a profusion of false ringlets, and gooseberry eyes, which were green in one light and grey in another.

Mr. Jorrocks's mother, who had long held a commission to get him a wife, had departed this life without executing it; and our friend soon finding himself going all wrong in his shirts and stocking-feet, and having then little time to go a-courting, just went, hand over head as it were, to a ball at the 'Horns' at Kennington Common, and drew the first woman that seemed inclined to make up to him, who chanced to be the now companion of his greatness.

(*Handley Cross*, 1843.)

THE COCKNEY SPORTSMAN

TWANG, twang, twang, went Mr. Jorrocks's horn, sometimes in full, sometimes in divided notes and half screeches. The hounds turn and make for the point. Governor, Adamant, Dexterous, and Judgment came first, then the body of the pack, followed by Benjamin at full gallop on Xerxes, with his face and hands all scratched and bleeding from the briars and brushwood, that Xerxes, bit in teeth, had borne him triumphantly through. Bang—the horse shot past Mr. Jorrocks, Benjamin screaming, yelling, and holding on by the mane, Xerxes doing with him just what he liked, and the hounds getting together and settling to the scent. 'My vig, wot a splitter!' cried Mr. Jorrocks in astonishment, as Xerxes took a high stone wall out of the cover in his stride, without disturbing the coping, but bringing Ben right on to his shoulder—'Hoff, for a fi' pun note! hoff for a guinea 'at to a Gossamer!' exclaimed Mr. Jorrocks, eyeing his whipper-in's efforts to regain the saddle. A friendly chuck of Xerxes's head assists his endeavours, and Ben scrambles back to his place. A gate on the left let Mr. Jorrocks out of cover, on to a good sound sward, which he prepared to take advantage of by getting Arterxerxes short by the head, rising in his stirrups, and hustling him along as hard as ever he could lay his legs to the ground. An open gate at the top fed the flame of his eagerness, and, not being afraid of the pace so long as there was no leaping, Jorrocks sent him spluttering through a swede turnip field as if it was pasture. Now sitting plum in his saddle, he gathered his great whip together, and proceeded to rib-roast Arterxerxes in the most summary manner, calling him a great, lurching, rolling, lumbering beggar, vowing that if he didn't lay himself out and go as he ought, he'd 'boil him when he got 'ome'. So he jerked and jagged, and kicked and spurred, and hit and held, making indifferent progress compared to his exertions. The exciting cry of hounds sounded in front and now passing on to a very heavy, roughly-ploughed upland, our master saw the hind-quarters of some half-dozen horses, the riders of which had been in the secret, disappearing through the high quick fence at the top.

'Dash my vig, 'eres an unawoidable leap, I do believe', said he to himself, as he neared the headland, and saw no way out of the field but over the fence—a boundary one; 'and a werry hawkward place it is too', added he, eyeing it intently, 'a yawnin' blind ditch, a hugly quick fence on the top, and may be, a plough or 'arrow turned teeth huppermost, on the far side.

'Oh, John Jorrocks, John Jorrocks, my good frind, I wishes you were well over with all my 'eart—terrible place, indeed! Give a guinea 'at to be at the far side', so saying, he dismounted, and pulling the snaffle-rein of the bridle over his horse's head, he knotted the lash of his ponderous whip to it, and very quietly slid down the ditch and climbed up the fence, 'who-a-ing' and crying to his horse to 'stand still', expecting every minute to have him atop of him. The taking-on place was wide, and two horses having gone over before, had done a little towards clearing the way, so having gained his equilibrium on the top, Mr. Jorrocks began jerking and coaxing Arterxerxes to induce him to follow, pulling at him much in the style of a school-boy who catches a log of wood in fishing.

'Come hup! my man', cried Mr. Jorrocks, coaxingly, jerking the rein; but Arterxerxes only stuck his great resolute fore legs in advance, and pulled the other way. 'Gently, old fellow!' cried he, 'gently, Arterxerxes, my bouy!' dropping his hand, so as to give him a little more line, and then trying what effect a jerk would have, in inducing him to do what he wanted. Still the horse stood with his great legs before him. He appeared to have no notion of leaping. Jorrocks began to wax angry. 'Dash my vig, you hugly brute!' he exclaimed, grinning with rage at the thoughts of the run he was losing, 'dash my vig, if you don't mind what you're arter, I'll get on your back and bury my spurs i' your sides. COME HUP! I say, YOU HUGLY BEAST!' roared he, giving a tremendous jerk of the rein, upon which the horse flew back; pulling Jorrocks downwards in the muddy ditch. Arterxerxes then threw up his heels and ran away, whip and all.

Meanwhile, our bagman played his part gallantly, running three-quarters of a ring, of three-quarters of a mile, chiefly in view, when, feeling exhausted, he threw himself into a furze-patch, near a farmyard, where Dauntless very soon had him by the back, but the smell of the aniseed, with which he had been plentifully rubbed, disgusting the hound, he chucked him in the air and let him fall back in the bush. Xerxes, who had borne Ben gallantly before the body of the pack, came tearing along, like a poodle with a monkey on his back, when, losing the cry of hounds, the horse suddenly stopped short, and off flew Benjamin beside the fox, who, all wild with fear and rage, seized Ben by the nose, who ran about with the fox hanging to him, yelling, 'Murder! murder! murder!' for hard life.

And to crown the day's disasters, when at length our fat friend got his horse and his hounds and his damaged Benjamin scraped together again, and re-entered Handley Cross, he was yelled at, and hooted, and rid coat! rid coat!-ed by the children, and made an object of unmerited ridicule by the fair but rather unfeeling portion of the populace.

'Lauk! here's an old chap been to Spilsby!' shouted Betty Lucas, the mangle-woman, on getting a view of his great mud-stained back.

'*Hoot*! he's always tumblin' off, that 'ard chap', responded Mrs. Hardbake, the itinerant lolly-pop seller, who was now waddling along with her tray before her.

'Sich old fellers have no business out a-huntin'!' observed Miss Rampling, the dressmaker, as she stood staring, bonnet-box on arm.

Then a marble-playing group of boys suspended operations to give Jorrocks three cheers; one, more forward than the rest, exclaiming, as he eyed Arterxerxes, 'A! what a shabby tail! A! what a shabby tail!'

Next as he passed the Barley-mow beer-shop, Mrs. Gallon, the landlady, who was nursing a child at the door, exclaimed across the street, to Blash, the barber's pretty but rather wordy wife—

'A—a—a! ar say Fanny!—old Fatty's had a fall!'

To which Mrs. Blash replied, with a scornful toss of her head, at our now admiring friend—

'*Hut*! he's always on his back, that old feller'.

'Not 'alf so often as you are, old gal!' retorted the now indignant Mr. Jorrocks, spurring on out of hearing.

(*Handley Cross*, 1843.)

THE MAN WHO CAME TO DINNER

THE room was a blaze of light. Countless compos swealed and simmered in massive gilt candelabras, while ground lamps of various forms lighted up the salmon-coloured walls, brightening the countenances of many ancestors, and exposing the dulness of the ill-cleaned plate.

The party having got shuffled into their places, the Rev. Jacob Jones said an elaborate grace, during which the company stood.

'I'll tell you a rum story about grace', observed Mr. Jorrocks to Mrs. Muleygrubs, as he settled himself into his seat, and spread his napkin over his knees. 'It 'appened at Croydon. The landlord o' the Grey'ound told a wise waiter, when a Duke axed him a question, always to say Grace. According, the Duke o' somebody, in changin' 'osses, popped his 'ead out o' the chay, and inquired wot o'clock it was. 'For what we're a-goin' to receive the Lord make us truly thankful', replied the waiter'.

Mrs. Muleygrubs either did not understand the story, or was too intent upon other things; at all events, Mr. Jorrocks's haw! haw! haw! was all that greeted its arrival.—But to dinner.

There were two soups—at least two plated tureens, one containing pea-soup, the other mutton-broth. Mr. Jorrocks said he didn't like the latter, it always reminded him of 'a cold in the 'ead'. The pea-soup he thought very like 'oss-gruel; that he kept to himself.

'Sherry or *My*-dearer?' inquired the stiff-necked boy, going round with a decanter in each hand, upsetting the soup-spoons, and dribbling the wine over people's hands.

While these were going round, the coachman and Mr. De

Green's boy entered with two dishes of fish. On removing the large plated covers, six pieces of skate and a large haddock made their appearance. Mr. Jorrocks's countenance fell five-and-twenty per cent., as he would say. He very soon despatched one of the six pieces of skate, and was just done in time to come in for the tail of the haddock.

'The Duke 'ill come on badly for fish, I'm thinkin',' said Mr. Jorrocks, eyeing the empty dishes as they were taken off.
'Oh, Marmaduke don't eat fish', replied Mrs. M.
'Oh, I doesn't mean your duke, but the Duke o' Rutland', rejoined Mr. Jorrocks.
Mrs. Muleygrubs didn't take.
'Nothing left for *Manners*, I mean, mum', explained Mr. Jorrocks, pointing to the empty dish.
Mrs. Muleygrubs smiled, because she thought she ought, though she didn't know why.
'Sherry or My-dearer, sir?' inquired the stiff-necked boy, going his round as before.
Mr. Jorrocks asked Mrs. Muleygrubs to take wine, and having satisfied himself that the sherry was bad, he took My-dearer, which was worse.
'Bad ticket, I fear', observed Mr. Jorrocks aloud to himself, smacking his lips. 'Have ye any swipes?'
'Sober-water and seltzer-water', replied the boy.
''Ang your sober-water!' growled Mr. Jorrocks.
'Are you a hard rider, Mr. Jorrocks?' now asked his hostess, still thinking anxiously of her dinner.
''Ardest in England, mum', replied our friend confidently, muttering aloud to himself, 'may say that, for I never goes off the 'ard road if I can 'elp it'.

After a long pause, during which the conversation gradually died out, a kick was heard at the door, which the stiff-necked foot-boy having replied to by opening, the other boy appeared, bearing a tray, followed by all the other flunkeys, each carrying a silver-covered dish.
'Come, *that's* more like the thing', said Mr. Jorrocks aloud to himself, eyeing the procession.

A large dish was placed under the host's nose, and another under that of Mrs. Muleygrubs.

'Roast beef and boiled turkey?' said Mr. Jorrocks to himself, half inclined to have a mental bet on the subject. 'May be saddle o' mutton and chickens', continued he, pursuing the speculation.

Four T. Cox Savory side-dishes, with silver rims and handles, next took places, and two silver-covered china centre dishes completed the arrangement.

'You've lots o' plate', observed Mr. Jorrocks to Mrs. Muleygrubs, glancing down the table.

'Can't do with less', replied the lady.

Stiffneck now proceeded to uncover, followed by his comrade. He began at his master, and, giving the steam-begrimed cover a flourish in the air, favoured his master's bald head with a hot shower-bath. Under pretence of admiring the pattern, Mr. Jorrocks had taken a peep under the side-dish before him, and seeing boiled turnips had settled that there was a round of beef at the bottom of the table. Spare ribs presented themselves to view. Mrs. Muleygrubs's dish held a degenerate turkey, so lean and lank that it looked as if it had been starved instead of fed. There was a reindeer tongue under one centre dish, and sausages under the other. Minced veal, forbidding-looking *rissoles*, stewed celery, and pigs' feet occupied the corner dishes.

'God bless us! what a dinner!' ejaculated Mr. Jorrocks, involuntarily.

'Game and black-puddings coming, isn't there, my dear?' inquired Mr. Muleygrubs of his wife.

'Yes, my dear', responded his obedient half.

' "*Murder most foul, as in the best it is;*
But this most foul, base, and unnattaral," '

muttered Mr. Jorrocks, running his fork through the breast of the unhappy turkey. 'Shall I give you a little *ding dong*?'

'It's turkey', observed the lady.

'True!' replied Mr. Jorrocks; '*ding dong's* French for turkey.'

'Are yours good hounds, Mr. Jorrocks?' now asked the lady, thinking how awkwardly he was carving.

'Best goin', mum!' replied our friend. 'Best goin', mum. The Belvoir may be 'andsomer, and the Quorn patienter under pressure, but for the real tear-'im and heat-'im qualities, there are none to compare wi' mine'. 'They're the bouys for making the foxes cry Capevi!' added our friend, with a broad grin of delight on his ruddy face.

'Indeed', mused the anxious lady, to whom our friend's comparisons were all gibberish.

'Shall I give anybody any turkey?' asked he, holding nearly half of it up on the fork preparatory to putting it on his own plate. Nobody claimed it, so our friend appropriated it.

Munch, munch, munch was then the order of the day. Conversation was very dull, and the pop and foam of a solitary bottle of 40$s.$ champagne, handed round much after the manner of liqueur, did little towards promoting it. Mr. Jorrocks was not the only person who wondered 'what had set him there'. Mrs. Muleygrubs attempted to relieve her agonies of anxiety by asking occasional questions of her guest.

'Are yours greyhounds, Mr. Jorrocks?' asked she with the greatest simplicity.

'No; greyhounds, no; what should put that i' your 'ead?' grunted our master with a frown of disgust; adding, as he gnawed away at the stringy drumstick, 'wouldn't take a greyhound in a gift.'

The turkey being only very so-so and the reindeer tongue rather worse, Mr. Jorrocks did not feel disposed to renew his acquaintance with either, and placing his knife and fork resignedly on his plate, determined to take his chance of the future. He remembered that in France the substantials sometimes did not come till late on.

Stiffneck, seeing his idleness, was presently at him with the dish of mince.

Mr. Jorrocks eyed it suspiciously, and then stirred the sliced lemon and meat about with the spoon. He thought at first of taking some, then he thought he wouldn't, then he fixed he wouldn't. 'No', said he, 'no', motioning it away with his hand, 'no, I likes to chew my own meat'.

The *rissoles* were then candidates for his custom.

'Large marbles', observed Mr. Jorrocks aloud to himself—'large marbles', repeated he, as he at length succeeded in penetrating the hide of one with a spoon. 'Might as well eat lead', observed he aloud, sending them away too.

'I often thinks now', observed he, turning to his hostess, 'that it would be a good thing, mum, if folks would 'gree to give up these stupid make-believe side-dishes, mum, for nobody ever eats them, at least if they do they're sure to come off second best, for no cuk that ever was foaled can do justice to such a variety of wittles'.

'Oh! but, Mr. Jorrocks, how could you send up a dinner properly without them?' exclaimed the lady with mingled horror and astonishment.

'Properly without them, mum', repeated our master, coolly and deliberately; 'properly without them, mum—why that's jest wot I was meanin' ', continued he. 'You see your cuk 'as sich a multitude o' things to do, that it's hutterly unpossible for her to send them all in properly, so 'stead o' gettin' a few things well done, ye get a great many only badly done'.

'Indeed!' fumed the lady, bridling with contempt.

'The great Duke o' Wellington—no 'fence to the present one', observed Mr. Jorrocks, with a low bow to the table—'who, I'm proud to say, gets his tea o' me too, the great Duke o' Wellington, mum, used to say, mum, that the reason why one seldom got a hegg well biled was 'cause the cuk was always a doin' summut else at the same time, and that hobservation will apply purty well to most cuking hoperations'.

'Well, then, you'd have no plate on the table, I presume, Mr. Jorrocks?' observed the irascible lady.

'Plate on the table, mum—plate on the table, mum', repeated Mr. Jorrocks, with the same provoking prolixity, 'why I really doesn't know that plate on the table's of any great use. I minds the time when folks thought four silver side-dishes made gen'l'men on 'em, but since these Brummagem things turned hup, they go for a bit o' land—land's the ticket now', observed our master.

While this unpalatable conversation—unpalatable, at least, to our hostess—was going on, the first course was being removed; and a large, richly-ornamented cold game-pie made its appearance, which was placed before Mr. Muleygrubs.

'Large tart!' observed Mr. Jorrocks, eyeing it, thinking if he could help himself he might yet manage to make up his leeway: 'thought there was dark puddins comin',' observed he to his hostess.

'*Game* and *black* puddings', replied Mrs. Muleygrubs. 'This comes between courses always'.

'Never saw it afore', observed Mr. Jorrocks.

Mr. Marmaduke helped the pie very sparingly, just as he had seen the butler at Ongar Castle helping a *pâté de foie gras*; and putting as much on to a plate as would make about a mouthful and a half to each person, he sent Stiffneck round with a fork to let people help themselves. Fortunately for Mr. Jorrocks, neither Mr. nor Miss De Green, nor Miss Slowan nor Mrs. Muleygrubs took any, and the untouched plate coming to him, he very coolly seized the whole, while the foot-boy returned to the dismayed Mr. Muleygrubs for more. Putting a few more scraps on a plate, Mr. Muleygrubs sent off the pie, lest anyone should make a second attack.

By dint of plying a good knife and fork, our friend cleared his plate just as the second course made its appearance. This consisted of a brace of partridges guarding a diminutive snipe at the top, and three links of black-pudding at the bottom—stewed celery, potato chips, puffs, and tartlets forming the side-dishes.

'Humph!' grunted our friend, eyeing each dish as it was uncovered. 'Humph!' repeated he—'not much there—three shillins for the top dish, one for the bottom, and eighteen-pence, say, for the four sides—five and six—altogether—think I could do it for five. Howsomever, never mind', continued he, drawing the dish of game towards him. 'Anybody for a *gibier* as we say in France?' asked he, driving his fork into the breast of the plumpest of the partridges. Nobody closed with the offer.

'Pr'aps if you'd help it, and let it be handed round, someone will take some', suggested Mr. Muleygrubs.

'Well', said Mr. Jorrocks, 'I've no objection—none wotever—only, while these clumsey chaps o' yours are runnin' agin each other with it, the wittles are coolin'—that's all', said our master, placing half a partridge on a plate, and delivering it up to go on its travels. Thinking it cut well, Mr. Jorrocks placed the other half on his own plate, and taking a comprehensive sweep of the crumbs and bread sauce, proceeded to make sure of the share by eating a mouthful of it. He need not have been alarmed, for no one came for any, and he munched and cranched his portion in peace. He then ate the snipe almost at a bite.

'What will you take next, Mr. Jorrocks?' asked his hostess, disgusted at his rapacity.

'Thank 'ee, mum, thank 'ee', replied he, munching and clearing his mouth; 'thank 'ee, mum', added he, 'I'll take breath if you please, mum', added he, throwing himself back in his chair.

'Have you killed many hares, Mr. Jorrocks?' now asked his persevering hostess, who was sitting on thorns as she saw an entering dish of blancmange toppling to its fall.

'No, mum, none!' responded our master, vehemently, for he had an angry letter in his pocket from Captain Slaughter's keeper, complaining bitterly of the recent devastation of his hounds—a calamity that of course the keeper made the most of, inasmuch as friend Jorrocks, as usual, had forgotten to give him his 'tip'.

Our innocent hostess, however, never listened for the answer, for the blancmange having landed with the loss only of a corner tower, for it was in the castellated style of confectionery, she was now all anxiety to see what sort of a savoury omelette her drunken job-cook would furnish, to remove the black-puddings at the other end of the table.

During this interval, our master, having thrust his hands deep in the pockets of his canary-coloured shorts, reconnoitred the table to see who would either ask him to take wine, or who he should honour that way: but not seeing any very prepossessing phiz, and recollecting that Mrs. J. had told him the good old-fashioned custom was 'wulgar', he was about to help himself from a conveniently-placed decanter, when Stiffneck, seeing what he was at, darted at the decanter, and passing behind Mr. Jorrocks's chair, prepared to fill to his holding, when, missing his aim, he first sluiced our master's hand, and then shot a considerable quantity of sherry down his sleeve.

'Rot ye, ye great lumberin' beggar!' exclaimed Mr. Jorrocks, furiously indignant; 'rot ye, do ye think I'm like Miss Biffin, the unfortunate lady without harms or legs, that I can't 'elp myself?' continued he, dashing the wet out of his spoon cuff. 'Now, that's the wust o' your flunkey fellers', continued he in a milder tone to Mrs. Muleygrubs, as the laughter the exclamation caused had subsided. 'That's the wust o' your flunkey fellers', repeated he, mopping his arm; 'they know they'd never be fools enough to keep fellers to do nothin', and so they think they must be constantly meddlin'. Now, your women waiters are quite different', continued he; 'they only try for the useful, and not for the helegant. There's no flash about them. If they see a thing's under your nose, they let you reach it, and don't bring a dish that's steady on the table round at your back to tremble on their 'ands under your nose. Besides', added our master, 'you never see a bosky Batsay waiter, which is more than can be said of all dog 'un's'.

'But you surely couldn't expect ladies to be waited upon by women, Mr. Jorrocks', exclaimed his astonished hostess.

'I would, though', replied our master, firmly, with a jerk of his head—'I would, though—I'd not only 'ave them waited upon by women, but I'd have them served by women i' the shops, 'stead o' those nasty dandified counter-skippin' Jackanapes, wot set up their himperances in a way that makes one long to kick 'em'.

'How's that, Mr. Jorrocks?' asked the lady, with a smile at his ignorance.

' 'Ow's that, mum?' repeated our master—' 'Ow's that? Why, by makin' you run the gauntlet of p'raps a double row o' these poopies, one holloain' out—'Wot shall I show you to-day, mum?' Another, 'Now, mum! French merino embroidered robes!' A third, 'Paisley and French wove shawls, mum! or Russian sables! chinchillas! hermines!' or 'Wot's the next harticle, mum?' as if a woman's—I beg pardon—a lady's wants were never to be satisfied—Oh dear, and with Christmas a comin' on', shuddered Mr. Jorrocks, with upraised hands; 'wot a lot o' squabbles and contentions 'ill shortly be let loose upon the world—bonnets, ribbons, sarsnets, bombazeens, things that the poor paymasters expected 'ad come out of the 'ouse money, or been paid for long ago'.

While Mr. Jorrocks was monopolizing the attention of the company by the foregoing domestic 'lector' as it may be called, the denounced domestics were clearing away the sweets, and replacing them with a dish of red herrings, and a very strong-smelling, brown, soapy-looking cheese.

Our master, notwithstanding his efforts, being still in arrear with his appetite, thought to 'fill up the chinks', as he calls it, with cheese, so he took a liberal supply as the plate came round —nearly the half of it, in fact.

He very soon found out his mistake. It was strong, and salt, and leathery, very unlike what Paxton and Whitfield supplied him with.

'Good cheese! Mr. Jorrocks', exclaimed his host, up the table; 'good cheese, eh?'

'Humph!' grunted our master, munching languidly at it.

'Excellent cheese, don't you think so, Mr. Jorrocks?' asked his host boldly.

'C-h-i-l-dren', drawled our master, pushing away his unfinished plate, 'would eat any q-u-a-a-n-tity of it'.

The clearing of the table helped to conceal the ill-suppressed titter of the company.

(Handley Cross, 1843.)

CLASSIC DISCOVERY

THE day continued hazy, but still no rain fell. Junks, however, persisted in his admonitions, and Mr. Jorrocks felt so certain it would rain, that he had Pigg into the parlour in the evening to make arrangements for the morning. Mrs. Jorrocks, Belinda, and Stobbs had gone out to tea, and Mr. Jorrocks was left all alone.

Master and man had an anxious confabulation. Mr. Jorrocks was all for Pinch-me-near; while Pigg recommended Hewtimber Forest.

Of course Jorrocks carried his point.

About nine Betsey brought the supper-tray, and Jorrocks would treat Pigg to a glass of brandy-and-water. One glass led to another, and they had a strong talk about hunting. They drank each other's healths, then the healths of the hounds.

'I'll give you old Priestess' good 'ealth!' exclaimed Mr. Jorrocks, holding up his glass. 'Fine old betch, with her tan eye-brows—thinks I never saw a better 'ound—wise as a Christian!' Pigg proposed Manager. Mr. Jorrocks gave Ravager. Pigg gave Lavender, and they drank Mercury, and Affable, and Crowner, and Lousey, and Mountebank, and Milliner—almost all the pack, in short, each in turn being best. A, what a dog one was to find a fox. A, what a dog another was to drive a scent.

The fire began to hiss, and Mr. Jorrocks felt confident his prophecy was about to be fulfilled. 'Look out of the winder, James, and see wot'un a night it is', said he to Pigg, giving the log a stir, to ascertain that the hiss didn't proceed from any dampness in the wood.

James staggered up, and after a momentary grope about the

room—for they were sitting without candles—exclaimed, 'Hellish dark, and smells of cheese!'

'Smells o' cheese!' repeated Mr. Jorrocks, looking round in astonishment; 'smells o' cheese!—vy, man, you've got your nob i' the cupboard—this be the vinder'.

(*Handley Cross*, 1843.)

ENTER MR. SPONGE

It was a murky October day that the hero of our tale, Mr. Sponge, or Soapey Sponge, as his good-natured friends call him, was seen mizzling along Oxford Street, wending his way to the West. Not that there was anything unusual in Sponge being seen in Oxford Street for when in town his daily perambulations consist of a circuit, commencing from the Bantam Hotel in Bond Street into Piccadilly, through Leicester Square, and so on to Aldridge's, in St. Martin's Lane, thence by Moore's sporting-print-shop, and on through some of those ambiguous and tortuous streets that, appearing to lead all ways at once and none in particular, land the explorer, sooner or later, on the south side of Oxford Street.

Oxford Street acts to the north part of London what the Strand does to the south; it is sure to bring one up, sooner or later. A man can hardly get over either of them without knowing it. Well, Soapey having got into Oxford Street, would make his way at a squarey, in-kneed, duck-toed, sort of pace, regulated by the bonnets, the vehicles, and the equestrians he met to criticize; for of women, vehicles, and horses, he had voted himself a consummate judge. Indeed he had fully established in his own mind that Kiddey Downey and he were the only men in London who *really* knew anything about horses, and fully impressed with that conviction, he would halt, and stand, and stare, in a way that with any other man would have been considered impertinent. Perhaps it was impertinent in Soapey—we don't mean to say it wasn't—but he had done it so long, and was of so sporting â gait and cut, that he felt himself somewhat privileged. Moreover, the majority of horsemen are so satisfied with the animals they bestride, that they cock up their jibs and ride along with a 'find any fault with either me or my horse, if you can' sort of air.

Thus Mr. Sponge proceeded leisurely along, now nodding to this man, now jerking his elbow to that, now smiling on a phaeton, now sneering at a 'bus. If he did not look in at Shackell's, or Bartley's, or any of the dealers on the line, he was always to be found about half-past five at Cumberland Gate, from whence he would strike leisurely down the Park, and after coming to a long check at Rotten Row rails, from whence he would pass all the cavalry in the Park in review, he would wend his way back to the Bantam, much in the style he had come. This was his summer proceeding.

Mr. Sponge had pursued this enterprising life for some 'seasons'—ten at least—and supposing him to have begun at twenty or one-and-twenty, he would be about thirty at the time we have the pleasure of introducing him to our readers—a period of life at which men begin to suspect they were not quite so wise at twenty as they thought. Not that Mr. Sponge had any particular indiscretions to reflect upon, for he was tolerably sharp, but he felt that he might have made better use of his time, which may be shortly described as having been spent in hunting all the winter, and in talking about it all the summer. With this popular sport he combined the diversion of fortune-hunting, though we are concerned to say that his success, up to the period of our introduction, had not been commensurate with his deserts. Let us, however, hope that brighter days are about to dawn upon him.

Having now introduced our hero to our male and female friends, under his interesting pursuits of fox and fortune-hunter, it becomes us to say a few words as to his qualifications for carrying them on.

Mr. Sponge was a good-looking, rather vulgar-looking man. At a distance—say ten yards—his height, figure, and carriage gave him somewhat of a commanding appearance, but this was rather marred by a jerky, twitchy, uneasy sort of air, that too plainly showed he was not the natural, or what the lower orders call the *real* gentleman. Not that Sponge was shy. Far from it. He never hesitated about offering to a lady, after a three days' acquaintance, or in asking a gentleman to take him a horse in over-night, with whom he might chance to come in contact in the hunting-field. And he did it all in such a cool, off-hand, matter-of-course sort of way, that people who would have stared with astonishment if anybody else had hinted at such a proposal, really seemed to come into the humour and spirit of the thing, and to look upon it rather as a matter of course than otherwise. Then his dexterity in getting into people's houses was only equalled by the difficulty of getting him out again, but this we must waive for the present in favour of his portraiture.

In height, Mr. Sponge was about the middle size—five feet eleven or so—with a well borne up, not badly shaped, closely cropped oval head, a tolerably good, but somewhat receding forehead, bright hazel eyes, Roman nose, with carefully tended whiskers, reaching the corners of a well-formed mouth, and thence descending in semicircles into a vast expanse of hair beneath the chin.

Having mentioned Mr. Sponge's groomy gait and horsey propensities, it was almost needless to say, that his dress was in the sporting style—you saw what he was by his clothes. Every article seemed to be made to defy the utmost rigour of the elements. His hat (Lincoln and Bennett) was hard and heavy. It sounded upon an entrance-hall table like a drum. A little magical loop in the lining explained the cause of its weight. Somehow, his hats were never either old or new—not that he bought them second-hand, but when he bought a new one he took its 'long-coat' off, as he called it, with a singeing lamp, and made it look as if it had undergone a few probationary showers.

When a good London hat recedes to a certain point, it gets no worse; it is not like a country-made thing that keeps going and going until it declines into a thing with no sort of resemblance to its original self. Barring its weight and hardness, the Sponge hat had no particular character apart from the Sponge head. It was not one of those punty ovals or Cheshire-cheese flats, or curly-sided things that enables one to say who is in a house and who is not, by a glance at the hats in the entrance, but it was just a quiet, round hat, without anything remarkable, either in the binding, the lining, or the band, still it was a very becoming hat when Sponge had it on. There is a great deal of character in hats. We have seen hats that bring the owners to the recollection far more forcibly than the generality of portraits. But to our hero.

That there may be a dandified simplicity in dress, is exemplified every day by our friends the Quakers, who adorn their beautiful brown Saxony coats with little inside velvet collars and fancy silk buttons, and even the severe order of sporting costume adopted by our friend Mr. Sponge, is not devoid of capability in the way of tasteful adaptation. This Mr. Sponge chiefly showed in promoting a resemblance between his neckcloths and waistcoats. Thus, if he wore a cream-coloured cravat, he would have a buff-coloured waistcoat, if a striped waistcoat, then the starcher would be imbued with somewhat of the same colour and pattern. The ties of these varied with their texture. The silk ones terminated in a sort of coaching fold, and were secured by a golden fox-head pin, while the striped starchers, with the aid of a pin on each side, just made a neat, unpretending tie in the middle, a sort of miniature of the flagrant, flyaway, Mile-End ones of aspiring youth of the present day. His coats were of the single-breasted cut-away order, with pockets outside, and generally either Oxford mixture or some dark colour, that required you to place him in a favourable light to say what it was.

His waistcoats, of course, were of the most correct form and material, generally either pale buff, or buff with a narrow stripe, similar to the undress vests of the servants of the Royal Family, only with the pattern run across instead of lengthways, as those worthies mostly have theirs, and made with good honest step-collars, instead of the make-believe roll collars they sometimes convert their upright ones into. When in deep thought, calculating, perhaps, the value of a passing horse, or considering whether he should have beefsteaks or lamb chops for dinner, Sponge's thumbs would rest in the arm-holes of his waistcoat; in which easy, but not very elegant, attitude, he would sometimes stand until all trace of the idea that elevated them had passed away from his mind.

In the trouser line he adhered to the close-fitting costume of former days; and many were the trials, the easings, and the alterings, ere he got a pair exactly to his mind. Many were the customers who turned away on seeing his manly figure filling the swing mirror in 'Snip and Sneiders', a monopoly that some tradesmen might object to, only Mr. Sponge's trousers being admitted to be perfect 'triumphs of the art', the more such a walking advertisement was seen in the shop the better. Indeed, we believe it would have been worth Snip and Co.'s while to have let him have them for nothing. They were easy without being tight, or rather they looked tight without being so; there wasn't a bag, a wrinkle, or a crease that there shouldn't be, and strong and storm-defying as they seemed, they were yet as soft and as supple as a lady's glove. They looked more as if his legs had been blown in them than as if such irreproachable garments were the work of men's hands. Many were the nudges, and many the 'look at this chap's trousers', that were given by ambitious men emulous of his appearance as he passed along, and many were the turnings round to examine their faultless fall upon his radiant boot. The boots, perhaps, might come in for a little of the glory, for they were beautifully soft and cool-looking to the foot, easy without [being] loose, and he preserved the lustre of their polish, even up to the last moment of his walk. There never was a better man for getting through dirt, either on foot or horse-back, than our friend.

To the frequenters of 'the corner', it were almost superfluous to mention that he is a constant attendant. He has several volumes of 'catalogues', with the prices the horses have brought set down in the margins, and has a rare knack of recognizing old friends, altered, disguised, or disfigured as they may be—'I've seen that rip before', he will say, with a knowing shake of the head, as some woe-begone devil goes, best leg foremost, up to the hammer, or, 'What! is that old beast back? why he's here every day'. No man can impose upon Soapey with a horse. He can detect the rough-coated plausibilities of the straw-yard, equally with the metamorphosis of the clipper or singer. His practised eye is not to be imposed upon either by the blandishments of the bang-tail, or the bereavements of the dock. Tattersall will hail him from his rostrum with—'Here's a horse will suit you, Mr. Sponge! cheap, good, and handsome! come and buy him'.

But it is needless describing him here, for every out-of-place groom and dog-stealer's man knows him by sight.

(*Mr. Sponge's Sporting Tour*, 1853.)

THE SPONGE CIGAR AND BETTING-ROOMS

ARRIVED in the great metropolis, Facey's earliest visit was to the well-known Soapey Sponge, in Jermyn Street, St. James's. Soapey in bygone days had been a guest of Facey's, and had lost a certain sum of 'sivin pun ten' to him at Blind Hookey, which no amount of coaxing and bullying had ever been able to extract from him. Indeed, latterly the letters had been returned to Facey through the dead-letter office. This was not to be borne, and Facey was now more than ever determined to have his dues, or to know the 'reason why'. We may mention that Soapey, on his marriage with the fascinating actress, Miss Lucy Glitters, had set up a cigar and betting shop, a happy combination, that promised extremely well at the outset, but an unfeeling legislature, regardless of vested interests, had presently interposed, and put a stop to the betting department. So Soapey had to extinguish his lists, and Lucy and he were reduced to the profits of the cigar shop alone,—'WHOLESALE, RETAIL, AND FOR EXPORTATION', as the circular brass front and window blind announced.

Now, though Lucy's attractions were great, and though she never sold even one of her hay-and-brown-paper cigars under sixpence, or ever gave change for a shilling, still Soapey and she could not make both ends meet; and when poverty comes in at the door, love will fly out of even a glittering cigar-shop window. So it was with the Sponges. Deprived of his betting recreation, Soapey took to idle and expensive habits; so true is the saying that

> '*Satan finds some mischief still*
> *For idle hands to do.*'

He frequented casinoes and billiard-rooms, danced at Cremorne, and often did not come home till daylight did appear. All this went sadly against the till; and the rent and the rates and taxes, to say nothing of the tradesmen's bills, were more difficult to collect on each succeeding quarter. With this falling fortune friend Facey arrived in town to further complicate disasters. He took two-pennyworth of Citizen 'bus from Lisson Grove as far as Piccadilly Circus, and then, either not knowing the country, or with a view of drawing up wind, threw himself into cover at the St. James's Street end of Jermyn Street, instead of at the Haymarket end, where one would have thought his natural genius would have suggested the Sponges would be found. To be sure he had not been much about town; Oncle Gilroy, for obvious reasons, having kept him as much as he could in the country. As we said before, it being the winter season, when day is much the same as night in London, Facey lounged leisurely along the gaslit street, one roguish eye reading the names and callings of the shops on his left, the other raking the opposite side of the way; but though he drew along slowly and carefully, examining as well the doors as the windows, no Soapey sign, no cigar warehouse, greeted his optics. Fish, books, boxes, bacon, boots, shoes, everything but Sponges.

So he came upon the 'bus-crowded Regent Street, not having had a whiff of a cigar save from the passers-by. There then he stood at the corner of the street biting his nails, lost in astonishment at the result. 'Reg'lar do', muttered he; 'beggar's bolted', looking back on the long vista of lamps he had passed. 'Well, that's a nice go', said he; 'always thought that fellow was a sharper'. Just then an unhandsome Hansom came splashing and tearing along the way he had come, and dashing across Regent Street pursued the continuous route beyond.

'May as well cast across here', said Facey to himself, picking his way over the muddy street, taking care of his buttoned boots as he went. His sagacity was rewarded by reading 'Jermyn Street' on the opposite wall. 'For-rard! for-rard!' he cheered himself, thinking the cigar-shop scent improved as he went. Indeed he quickly came upon a baccy shop, green door, red blinds, all indicative of a find, for no sooner does one tradesman get well-established than another comes as near as he can get to pick away part of his custom.

Just then Facey's keen eye caught sight of two little over-dressed snobs stopping suddenly at a radiant shop window a few paces further on, and advancing stealthily along, as if going up with his gun to a point, the words 'Devilish 'andsome' fell upon his ear. Looking over their shoulders there appeared the familiar figure of Mrs. Sponge behind the counter. Mrs. Sponge, slightly advanced in *embonpoint* since he saw her, but still in the full bloom of womanly beauty. She was dressed in a semi-evening costume, low-necked lavender-coloured silk dress, with an imitation black Spanish mantilla thrown gracefully over her swan-like neck and drooping well-rounded shoulders. The glare of the gaslight illuminated her clear Italian-like complexion, and imparted a lustre to a light bandeau of brilliants that encircled her jet-black hair. Altogether she looked very bewitching. There was a great hairy fellow in the shop, as big as Facey, and better made, who kept laughing and talking, and 'Lucy'-ing Mrs. Sponge in the familiar way fools talk to women in bars and cigar rooms. The little snobs were rather kept at bay by the sight; not so friend Facey, who brushed past them and boldly entered the once famous 'Sponge Cigar and Betting-Rooms'. Lucy started with a half-suppressed shriek at the sight, for Romford at any time would have been formidable, but a black Romford was more than her nerves could bear. Added to this she knew who had returned the dunning letters, and feared the visit boded no good.

'Well, and how goes it?' said Facey, advancing, and tendering his great ungloved hand.

'Pretty well, thank you, Mr. Romford', replied Lucy, shaking hands with him.

'And how's the old boy?' asked Facey, meaning Soapey.

'He's pretty well, too, thank you', replied Mrs. Sponge.

'At home?' asked Facey, with an air of indifference.

'Well—no—' hesitated Lucy, 'he's just gone out to his drill. He is one of the West Middlesex.' (He was upstairs dressing to go to the billiard-room.)

The hairy monster seeing he was superseded presently took his departure, and the little snobs having passed on, the two were left together; so Facey taking a chair planted himself just opposite the door, as well to stare at her as to stem the tide of further custom. It was lucky he did, for Sponge coming downstairs peeped through the dun-hole of the little retiring room, and recognizing his great shoulders and backward-growing whiskers, beat a retreat and stole out the back way.

'And may I ask who you are in mourning for?' inquired Lucy, as soon as the first rush of politeness was over.

'Oh, me Oncle Gilroy', replied Facey.

'Gone at last, is he', said Lucy, who recollected to have heard about him.

'Gone at last', assented Facey, with a downward nod.

'Well, and I hope he's left you something 'andsome', observed Lucy.

'Leave! Oh, bless you, I never expected nothin' from him. He had a wife and ever so many bairns.'

'You don't say so!' exclaimed Lucy, clasping her beautiful hands; I always understood he was a bachelor. Well, Mr. S. *will* be astonished when he hears that', added she, turning her lustrous darkly-fringed eyes up to the ceiling.

'Fact, however', said Facey significantly.

'You surprise me', said Lucy, fearing the little debt would not be wiped off. 'Well', continued she, 'it's lucky for those that can do without'.

'Ah, that's another matter', muttered Facey, who saw how it bore on the sivin pun ten. 'Money's always acceptable', continued he, looking round the shining shop and wondering if he would ever get paid. There seemed plenty of stock, provided the barrels and canisters were not all dummies. How would it do to take it out in kind? Better get money if he could, thought he. Facey then applied himself to sounding Lucy as to where Sponge was likely to be found. Oh, he would be sure to find him at any time; could scarcely come wrong. He hadn't been gone five minutes when Mr. Romford came. Would be so vexed when he returned to find he'd missed him. Facey rather doubted this latter assertion, and was half inclined to ask why Soapey had not answered his letters, but Lucy being too pretty to have any words with, and appearing to believe what she said, he pretended he did too, and shortly afterwards left to get a beefsteak dinner at the Blue Posts in Cork Street.

(*Mr. Facey Romford's Hounds*, 1865.)

EARLY-VICTORIAN EXQUISITE, MALE

'Now, Bray, don't make yourself such a swell', said young Lord Aubrey, entering the Marquis's room, who, with the aid of his valet, was settling himself into one of Jackson's particulars, blue coat, velvet collar and cuffs, silk facings and linings, with Windsor buttons. Nature meant the Marquis for a girl, and a very pretty one he would have made. He had a beautiful pink and white complexion, hair parted down the middle of his head, and falling in ringlets about his ears, blue eyes, Grecian nose, simpering mouth, with a dimple on each side, very regular pearly teeth, an incipient moustache on his upper lip, and a very incipient imperial on a very pretty unshaved chin. In stature he was about the middle height, five feet ten or so, thin, with a deal of action in his legs and backbone; indeed, he had a considerable cross of the dancing-master in him, and was considered one of the best 'goers' at Almack's or the Palace. In short, he was a pretty Jemmy Jessamy sort of fellow.

Now, this sort of man is generally desperately disliked by their own sex, particularly by the hirsute, rasping bull-finching breed of fox-hunters; and just in proportion as men are abused by each other, they are petted and praised by the women—particularly if they are marquises, and *in the market*.

Accordingly, our hero stood as an 'A1' lady-killer in London; and that being the case, our readers may imagine what a desperate man he would be in the country. Indeed, these sort of fellows ought not to be allowed to go about unmuzzled (that is to say, without a wife), for country girls are monstrous inflammatory, and having little choice beyond the curate and the apothecary's apprentice, are ready to worry anything in the shape of a man—to say nothing of a lord—a handsome Marquis beyond all conception. Then the greasy novels put such notions into their heads. We really believe they think the great people go into the country for wives, just as the Cockneys go to Kensington for strawberries and cabbages; and there is nothing of the sort to be had in London. Unfortunately for rural belles, London beaux look upon them in quite a different light. They consider them a sort of strop to keep the razor of their palaverment fresh against the return of another London season, and think they may go any length short of absolutely offering; and that the girls wash the slates of their memories just as they wash their own on passing Hyde Park, down Portland Place, or by the Elephant and Castle, on their way back to town. The Marquis of Bray was just one of this sort. He knew perfectly well the Duke would no more think of letting him marry anything below a Duke's daughter, than he would think of sending him off for a trip in one of Mr. Henson's air carriages; and being well assured of that fact, he thought the girls must know it also, and would just take his small talk for what it was meant. Moreover, the Marquis having had the unspeakable misfortune of being brought up at home, had conceived the not at all unnatural idea that the world was chiefly made for him, and that he might do whatever he liked with impunity. No greater misfortune surely can befall a young man than such an education; and lucky it is that so few of them get it. Eton knocks and Eton kicks save many a 'terrible high-bred' lad (as the Epsom race-list sellers describe the horses) from ruin.

But we must get the Marquis downstairs. Behold him, then, in his blue coat aforesaid, with a delicate bouquet in the buttonhole—a most elaborately-tied white cravat, the folds of the tie nestling among six small point-lace frills of an exquisitely embroidered lawn shirt front over a pink silk under-waistcoat, and diamond studs of immense value, chained with Lilliputian chains—his waistcoat of cerulean blue satin, worked with heart's-ease, buttoned with buttons of enormous bloodstones, the surface of the waistcoat traversed with Venetian chains and diminutive seals—pink silk stockings, and pumps—gliding into the drawing-room, with an airy noiseless tread, and a highly scented, much-embroidered, lace-trimmed handkerchief in his hand. How he bowed! how he smiled! how he showed his teeth! He was so d—d polite, you'd have thought he'd got among a party of emperors, instead of among all the John Browns of the neighbourhood. Then the old Duke, like all blunderheaded men, being monstrously afraid lest his son should make mistakes, must needs take him in hand, and introduce him to those he didn't know. 'Jeems, my dear!' cried he, as the elastic back began to slacken in its salaams round the awe-stricken circle, 'come here, and let me introduce you to our excellent friend, Mr. Jorrocks, who's been kind enough to come all the way from —from—from—to dine with us'.

'To dine and *stay all night,* your Greece', observed Mr. Jorrocks to the Duke, letting fall his coat laps, preparatory to offering his hand to the Marquis.

The Marquis bowed and grinned, and laid his hand upon his heart, as if perfectly overcome by the honour—proudest moment of his life!

'Where I dine I sleep, and where I sleep I breakfast, your Greece', observed Mr. Jorrocks, resuming his position, finding it impossible to compete with the Marquis in bows.

(Hillingdon Hall, 1845.)

EARLY-VICTORIAN
EXQUISITE, FEMALE

WHEN the mortified Miss de Glancey returned to her lodgings at Mrs. Sarsnet the milliner's, in Verbena Crescent, she bid Mrs. Roseworth good night, and dismissing her little French maid to bed, proceeded to her own apartment, where, with the united aid of a chamber and two toilette-table candles, she instituted a most rigid examination, as well of her features as her figure, in her own hand-mirror and the various glasses of the room, and satisfied herself that neither her looks nor her dress were anyway in fault for the indifference with which she had been received. Indeed, though she might perhaps be a little partial, she thought she never saw herself looking better, and certainly her dress was as stylish and looming as any in the ballroom.

Those points being satisfactorily settled, she next unclasped the single row of large pearls that fastened the bunch of scarlet geraniums into her silken brown hair; and taking them off her exquisitely modelled head, laid them beside her massive scarlet geranium bouquet and delicate kid gloves upon the toilette-table. She then stirred the fire; and wheeling the easy-chair round to the front of it, took the eight hundred yards of tulle deliberately in either hand and sunk despondingly into the depths of the chair, with its ample folds before her. Drawing her dress up a little in front, she placed her taper white-satined feet on the low green fender, and burying her beautiful face in her lace-fringed kerchief, proceeded to take an undisturbed examination of what had occurred. How was it that she, in the full bloom of her beauty and the zenith of her experience, had failed in accomplishing what she used so easily to perform?

How was it that Captain Languisher seemed so cool, and that supercilious Miss eyed her with a side-long stare, that left its troubled mark behind, like the ripple of the water after a boat. And that boy Pringle, too, who ought to have been proud and flattered by her notice, instead of grinning about with those common country Misses? All this hurt and distressed our accomplished coquette, who was unused to indifference and mortification. Then from the present her mind reverted to the past; and stirring the fire, she recalled the glorious recollections of her many triumphs, beginning with her school-girl days, when the yeomanry officers used to smile at her as they met the girls out walking, until Miss Whippey restricted them to the garden during the eight days that the dangerous danglers were on duty. Next, how the triumph of her first offer was enhanced by the fact that she got her old opponent Sarah Snowball's lover from her—who, however, she quickly discarded for Captain Capers—who in turn yielded to Major Spankley. Then she thought how she kept the rich Mr. Acres, the gay Mr. Dicer, and the grave Mr. Woodhouse all in tow together, each thinking himself the happy man and the others the cat's-paw, until the rash Hotspur Smith exploded among them, and then suddenly dwindled from a millionaire into a mouse. Other names quickly followed, recalling the recollections of a successful career. At last she came to that dread, that fatal day, when, having exterminated Imperial John, and with the Peer well in hand, she was induced, much against her better judgment, to continue the chase, and lose all chance of becoming a Countess. Oh, what a day was that! She had long watched the noble Earl's increasing fervour, and marked his admiring eye, as she sat in the glow of beauty and the pride of equestrianism; and she felt quite sure, if the chase had ended at the check caused by the cattle-drover's dog he would have married her. Oh, that the run should ever have continued! Oh, that she should ever have been lured on to her certain destruction! Why didn't she leave well alone? And at the recollection of that sad, that watery day, she burst into tears and sobbed convulsively. Her feelings being thus relieved, and the fire about exhausted, she then got out of her crinoline and under the counterpane.

(*Ask Mamma*, 1858.)

ENGLISH MISS, c. 1845

As someone said of Talleyrand, that you might kick him behind without his countenance betraying a change, so a man might have kissed Emma Flather for half an hour without raising a blush on her cheeks. Indeed, she was a fine piece of animated statuary—and as cold withal. A provoking sort of girl. Not exactly pretty enough to fall in love with for her looks, and yet dangerous with her looks and blandishments combined. She was desperately enthusiastic; could assume raptures at the sight of a daisy, or weep o'er the fate of a fly in the slop-basin. Moreover, she had a smattering of accomplishments, could sing, and play, embroider, work worsteds, murder French and Italian, and had a knack of talking and pretending to a great deal more talent than she possessed. This taste for exaggeration she carried into other matters; she had a fine fertile imagination —frequently fancied herself a great heiress—talked of the beauty of her aunt's place in Dorsetshire—insinuated that she was to inherit it, with a vast number of other little self-enhancements, plainly showing that her *education* had not been neglected.

Emma was a curious mixture of high-mindedness and meanness—of feeling and insensibility. Full of enthusiasm and lofty sentiments—compassionate and tender beyond expression when it suited her purpose—she was, nevertheless, selfish and insensible to the last degree. Cold, calculating, and cunning, she had all the worldly-mindedness of a well-hackneyed woman of fifty —in short, of her mother. As the Frenchman said of his dog, 'she was well down to charge', and thoroughly appreciated the difference between an elder son and a younger. She would dismiss the latter at any moment that her mother hinted the probability of anything better. All this told in her favour, she acquired the character of a model of propriety, and Emma Flather was held up as a pattern girl for all young ladies to imitate. Of course, old mother Flather was extremely anxious to get her married, but not having fallen in with anything exactly to her mind, she had just flown her at minor game and checked her off under pretence of not being able to part with her dear girl.

(Hillingdon Hall, 1845.)

BEDROOM SCENE

THE Major having inducted his guest into one of those expensive articles of dining-room furniture, an easy chair—expensive, inasmuch as they cause a great consumption of candles, by sending their occupants to sleep,—now set a little round table between them, to which having transferred the biscuits and wine, he drew a duplicate chair to the fire for himself, and, sousing down in it, prepared for a *tete-a-tete* chat with our friend. He wanted to know what Lord Ladythorne said of him, to sound Billy, in fact, whether there was any chance of his making him a magistrate. He also wanted to find out how long Billy was going to stay in the country, and see whether there was any chance of selling him a horse; so he led up to the points, by calling upon Billy to fill up a bumper to the 'Merry haryers', observing casually, as he passed the bottle, that he had now kept them 'five-and-thirty years without a subscription, and was as much attached to the sport as ever'. This toast was followed by the foxhounds and Lord Ladythorne's health, which opened out a fine field for general dissertation and sounding, commencing with Mr. Boggledike, who, the Major not liking, of course, he condemned; and Mrs. Pringle having expressed an adverse opinion of him too, Billy adopted their ideas, and agreed that he was slow, and ought to be drafted.

Monsieur Jean Rougier having taken the general bearings of the family as he stood behind 'me lor Pringle's' chair, retired from active service on the coming in of the cheese, and proceeded to Billy's apartment, there to arrange the toilette table and see that everything was *comme il faut*. Billy's dirty boots, of course, he took downstairs to the Bumbler to clean, who, in turn, put them off upon Solomon.

Very smart everything in the room was. The contents of the gorgeous dressing-case were duly displayed on the fine white damask cloth that covered the rose-colour-lined muslin of the gracefully-fringed and festooned toilette cover, whose flowing drapery presented at once an effectual barrier to the legs, and formed an excellent repository for old crusts, envelopes, curl-papers, and general sweepings. Solid ivory hairbrushes, with tortoiseshell combs, cosmetics, curling fluids, oils and essences without end, mingled with the bijouterie and knick-nacks of the distinguished visitor. Having examined himself attentively in the glass, and spruced up his bristles with Billy's brushes, Jack then stirred the fire, extinguished the toilette-table candle, which he had lit on coming in, and produced a great blue blouse from the bottom drawer of the wardrobe, in which, having enveloped himself in order to prevent his fine clothes catching dust, he next crawled backwards under the bed. He had not laid there very long ere the opening and shutting of downstairs doors, with the ringing of a bell, was followed by the rustling of silks, and the light tread of airy steps hurrying along the passage, and stopping at the partially-opened door. Presently increased light in the apartment was succeeded by less rustle and tip-toe treads passing the bed, and up to the looking-glass. The self-inspection being over, candles were then flashed about the room in various directions; and Jack having now thrown all his energies into his ears, overheard the following hurried *sotto voce* exclamations:—

First Voice: 'Lauk! what a little dandy it is!'

Second Voice: 'Look, I say! look at his boots—one, two, three, four, five, six, seven, eight, nine, ten: ten pair, as I live, besides jacks and tops.'

First Voice: 'And shoes in proportion', the speaker running her candle along the line of various patterned shoes.

Second Voice (*advancing to the toilette-table*): 'Let's look at his studs. Wot an assortment! Wonder if those are diamonds or paste he has on.'

First Voice: 'Oh, *diamonds* to be sure' (with an emphasis on diamonds.) 'You don't s'pose such a little swell as that would wear paste. See! there's a pearl and diamond ring. Just fits me, I do declare', added she, trying it on.

Second Voice: 'What beautiful carbuncle pins!'
First Voice: 'Oh, what studs!'
Second Voice: 'Oh, what chains!'
First Voice: 'Oh, what pins!'
Second Voice: 'Oh, what a love of a ring!' And so the ladies continued, turning the articles hastily over. 'Oh, how happy he *must* be', sighed a languishing voice, as the inspection proceeded.

'See, here's his little silver shaving box', observed the first speaker, opening it.

'Wonder what *he* wants with a shaving box—got no more beard than I have', replied the other, taking up Billy's badger-hair shaving-brush, and applying it to her own pretty chin.

'Oh! smell what delicious perfume!' now exclaimed the discoverer of the shaving-box. 'Essence of Rondeletia, I do believe! No, extrait de millefleurs', added she, scenting her 'kerchief with some.

Then there was a hurried, frightened 'hush!' followed by a 'Take care that ugly man of his doesn't come.'

'Did you ever *see* such a monster!' ejaculated the other earnestly.

'Kept his horrid eyes fixed upon me the whole dinner', observed the first speaker.

'Frights they are', rejoined the other.

'He must keep him for a foil', suggested the first.

'Let's go or we'll be caught!' replied the alarmist; and forthwith the rustling of silks was resumed, the candles hurried past, and the ladies tripped softly out of the room, leaving the door ajar, with Jack under the bed, to digest their compliments at his leisure.

But Monsieur was too many for them. Miss had dropped her glove at the foot of the bed, which Jack found on emerging from his hiding-place, and waiting until he had the whole party re-assembled at tea, he walked majestically into the middle of the drawing-room with it extended on a plated tray, his 'horrid eyes' combining all the venom of a Frenchman with the *hauteur* of an Englishman, and inquired, in a loud and audible voice, 'Please, has any lady or shentleman lost its glo-o-ve?'

'Yes, I have!' replied Miss, hastily, who had been wondering where she had dropped it.

'Indeed, marm', replied Monsieur, bowing and presenting it to her on the tray, adding, in a still louder voice, 'I FOUND IT IN MONSIEUR PRINGLE'S BED-ROOM'. And Jack's flashing eye saw by the brightly colouring girls which were the offenders.

Very much shocked was Mamma at the announcement; and the young ladies were so put about, that they could scarcely compose themselves at the piano, while Miss Harriet's voice soared exultingly as she accompanied herself on the harp.

(Ask Mamma, 1858.)

PROFESSIONAL POLITICIAN

It was a fortunate day which secured to the Anti-Corn-Law League the services of Mr. William Bowker—fortunate to the League, for they gained an able and most unscrupulous coadjutor; and fortunate to Mr. William Bowker, for he had just lost the best part of his income by the demise of his old master, the celebrated Mr. Snarle, the great conveyancer of Lincoln's Inn.

Mr. William Bowker, or Bill, as he was familiarly called, was one of a large class of men about town, who make a very great show upon very slender means. Not that he made any equestrian or vehicular display, but in his person he was a most uncommon swell, gay and gaudy in his colours, glittering in his jewellery (or make believes), faultless in his hat, costly in his linen (or apologies), expensive in his gloves, and shining in his boots. Many a country cousin, and many a one again, has anxiously inquired of his London cicerone 'who that smart gentleman was', as Bill has strutted consequentially through the Park on a Sunday, swinging his cane, with the tassels of his Hessian boots tapping the signal of his approach.

Many a time Mr. Jorrocks and him have passed for lords as they rolled arm in arm through the Zoological or Kensington Gardens, *haw, haw, hawing* at each other's jokes, looking about at the girls and criticizing their feet and ankles. This latter, however, was in short-petticoat times.

Mr. Bowker was an extraordinary fellow; over head and ears in debt and difficulties, he was as light and gay as if he hadn't a care in the world. Not a new fashion came out but Bill immediately had it. If a flight of extraordinary neckcloths alighted in the mercers' windows, the next time you met Bill he was sure to have one on. All the rumbustical apologies for greatcoats that have inundated the town of late years had their turns on Bill's back. You seldom saw him twice in the same waistcoat. Variable as D'Orsay, and as gay in his colours. Moreover, there was a certain easy nonchalance about Bill, far different to the anxious eyeings and watchings of the generality of 'would-be' swells. He would salute a man immeasurably his superior, with perfect familiarity; offer his richly-ornamented gilt snuff-box, or poke him in the ribs with a smile and a wink, that plainly said, 'You and I have a secret between us'. His looks were in his favour—rosy and healthy, as though he had never known care or confinement, with wavy yellow locks, slightly streaked with grey, giving him the license of age over youngsters; while his jolly corpulency and plummy legs, filling his bright Hessian boots, had the appearance of belonging to some swell fox-hunter up at Long's or Limmer's or some of the tiger-traps, out for what they call a spree—*rouge et noir*, feathers, hot port, Clarence Gardens, and the Quadrant.

In the language of the sect, Bill had some breeding in him—by a lord, out of a lady's maid—and blood will tell in men as well as horses. Hence, whatever his difficulties, or whatever his situation, Bill always retained the easy composure of a well-bred man. His address was good, his manner easy, and his language pure. If fortune had neglected to supply him with the essentials, at all events it had not deprived him of the advantages of birth. He was about the gamest cock with the fewest feathers that ever flew.

Hundreds will exclaim on reading this sketch, 'Lord, I know that man as well as can be! Have seen him in the Park a thousand times'; and perhaps no one has caused more 'Who's that?' than our friend Mr. Bowker. Indeed, he was a sort of person that you couldn't overlook, any more than you could a peacock in a poultry yard, for there was a strut and a dazzle about him that almost provoked criticism. Of course Bowker was well known to his own set, but what's a man's own set in the great ocean of London society? Moreover, even in his own set he was an object of admiration, for he was friendly and jocose, and we don't believe there was a man among them but would rather have enhanced Bill's consequence than attempted to lower him by proclaiming him the clerk to a conveyancer, and keeper of a miserable tobacco shop in the miserable purlieus of Red Lion Square. Our readers, we dare say, will be anxious to know how Bill managed matters. We will tell them. *He lived by his wits.*

When old Snarle was in full practice, Bill's fees were considerable, and in those days he was nothing but the 'thorough varmint and the real swell'. As soon as Chambers closed, he repaired, full dress, to a theatre, attended a 'free and easy', or some convivial society. Here his jolly good humour ensured him a hearty reception, and the landlords of the houses were too happy to hand him anything he called for in return for the amusement he afforded to their customers. He could sing, or he could talk, or he could dance, or he could conjure, lie through thick and thin—in short, do everything that's wanted at this sort of place. He was in with the players too, and had the *entrée* of most of the minor theatres about London. At these he might be seen in the front row of the stage boxes, dressed out in imitation of some of the fat swells in the 'omnibus', his elbow resting on a huge bamboo, with a large 'Dollond' in his primrose-kidded hand. There he was with the critic. Not the noisy, boisterous, self-proclaiming *claqueur*, but the gentle, irresistible leader, whose soft plaudits brought forth the thunder of the pit and gallery. He had some taste for acting, and we have read some neatish *critiques* attributed to him in the *Morning Herald and Advertiser*. This sort of society brought him, of course, a good deal among actresses, and we have heard that several of his 'How d'ye do?' great acquaintance arose out of little delicate arrangements that he had the felicity of bringing about. This, however, we don't vouch for; we will therefore thank our readers not to 'quote us' on this point.

But to the 'baccy' shop.

As fees fell off, Bill set up a snuff and cigar shop, and he who had amused so many, sought for the favours of the fumigating public. But Bill had a great mind. He did not stoop to the humble-mindedness of appearing as a little tobacconist, but leapt all at once into the station of a merchant, and advertised his miserable domicile as BOWKER AND CO.'S WHOLESALE SNUFF AND TOBACCO WAREHOUSE—THE TRADE SUPPLIED.

Whether this latter announcement had the effect of keeping off customers—people perhaps supposing they could not get less than a waggon-load of baccy at a time,—or whether Eagle Street is too little of a thoroughfare, or not sufficiently inviting in its appearance, or whether there were too many Bowker & Co.'s in the trade already, we know not; but certain it is, no wholesale customer ever cast up, and most of the retail ones were what Bill touted himself or were brought by his friends. The situation, we take it, must have been the thing; not that we mean to say anything unhandsome of Eagle Street, but we cannot account for the bad success of Bowker & Co.'s establishment upon any other grounds than that the neighbouring shops were not attractive, and a good deal of a tobacconist's trade consists of what is called 'chance custom'. Doors with half-a-dozen bell-pulls in each post, denoting half-a-dozen families in the house, coal and cabbage sheds united, those mysterious, police-inviting bazaars, denominated 'marine stores,' with milk shops, corn chandlers, furniture warehouses, and pawnbrokers commingled, do not add much to the appearance of any street, and certainly Eagle Street has nothing to lose in the way of attraction.

Yes, the situation must have been the thing, for if any one will take the trouble of walking through the thoroughfares, and casting their eyes into the brilliantly-illuminated 'divans', they will see men, without a tithe part of Mr. Bowker's ready wit and humour, handing the cigars over the counter as fast as they can fumble them, with women immeasurably Mrs. Bowker's inferior, riveting men with their charms, and sending them away by the score every night with the full conviction that they are desperately in love with them all, and only wanting to get rid of the other chaps to tell them so. *That*, we take it, is the grand secret of the baccy shop. Keep up the delusion, and you keep up your customers, but then you must have a bumper at starting. There's the advantage of a thoroughfare. Fool No. 2 sees Fool No. 1 smoking and making eyes at a woman, and in he goes to see what she's like. She's equally affable with him, and while both are striving to do the agreeable in comes No. 3 on a like errand—4, 5, 6, 7, 8, 9, 10—legion, in fact, quickly follow, and they all go on eyeing and fumigating, as jealous of each other as ever they can be, until the smoke obscures their vision, and they leave, each with the determination of seeing what they can do single-handed next night. The shop is then established.

Mrs. Bowker, when Bill set up, was a fine, big, dashing woman, with as good a foot and ankle as any in London. She was then on the stage at the Coburg, but marrying Bill for the purpose of getting off it, he found to his sorrow that she was likely to be a dead weight, instead of an assistance in housekeeping and theatrical society, which it was then his ambition to enter. Still there were her looks—a clear Italian complexion, large richly-fringed dark eyes, cork-screwy ringlets, swan-like neck and ample bust; and what with gaslight, and the tinsel of a theatrical wardrobe, Bill hoped to turn his better half to some account in the way of decoy duck at a cigar shop. Mrs. Bowker, however, took badly to it. She was above it, in fact, and instead of sitting to display her charms in the gaslight, she was generally sipping brandy-and-water, and reading greasy novels on a sofa in the back-shop. Miss Susan Slummers, her sister, also an actress and a fine handsome girl too, was shortly afterwards added to the family circle; and certainly, if wit and beauty can command success in the baccy line, Mr. Bowker had every reason to expect it. Still, as we said before, we grieve to say it did not come; and debt, and duns, and difficulties soon beset Bill's path of life in most alarming profusion.

Our old friend, Mr. Jorrocks, as kind-hearted and liberal a man as ever stuffed big calves into top-boots, long stood his friend—so long indeed, that the worthy old gentleman had ceased entering Bill's obligations in his books—and many people trusted Bill on the strength of the intimacy, who would never have let him into their debt upon the faith of any of his own palaverments. Not that he was a bad hand in that line, but they had had too much of it. In short, Bill was better known than trusted.

Thus then matters stood at the time of Bill's enlistment in the League. Old Snarle was dead. The dwindling fees were done. To begin brushing coats and cleaning boots for a new man, in hopes of seeing him rise in the profession, was out of the question to a man with Bill's ideas, and at his time of life. The cigar-shop did nothing. Mrs. Bowker did a good deal in the brandy-and-water way. House rent was due—their first floor lodger had left them. Gas rent was in arrear—water ditto—and poors' rate collecting. Income-tax, we needn't say, he was exempt from.

Mr. Jorrocks had retired into the country, and though he had never turned a deaf ear to any of Bill's representations or petitions, still our worthy tobacconist could not help feeling that without the aid of the emollient blarney wherewith to pave the way in jolly half-seas-over intimacy, the ominous 'no effects' might some day be returned to his epistolatory requisitions, and then what *was* to become of him?

—The law and Mr. Commissioner Fonblanque only knew!

Having now introduced Mr. Bowker, we will let his correspondence with Mr. Jorrocks speak for his situation and arrangements.

'Eagle Street, Red Lion Square.
'Honoured Sir,

'You'll be glad to hear that your old friend Bill has lit on his legs at last. High time he did, for I really think I was never so nearly stumpt in my life. Old Snarle, as you'll have heard, has cut his stick. Poor old bitch! Yet let it not be as our great master says:—

' "*—the evil that men do lives after them;*
The good is oft buried with their bones."

'Snarle had his faults, and so have we all, but for 'parties in a hurry', there never was a quicker hand at a settlement. May his new settlement be to his liking!

'T'other night, as I was sitting in my back-shop uncommonly spooney, reflecting on the uncertainty of life, and the certainty of the tax-gatherer calling in the morning, a mysterious, big black-whiskered, beetle-browed stranger entered the shop, and asked to have a word with me in private. As soon as we had coalesced behind the scenes, "Mr. Bowker", said he, taking off his broad-brimmed hat and gloves, laying them on the table, and sitting down on the sofa, as if he meant to be comfortable.

' "You don't know me?"

' "Why, you have the advantage of me", said I.

' "Well," said he, "I come to advantage you."

' "Glad of it", says I, adding aside, "wonder if it's Joseph Ady"

' "You are to be depended upon?" said he, after a pause.
' "Close as wax", said I.
' "Well, then", said he, "you have heard of the great National Anti-Corn-Law League?"
' "I have seen their advertising machine", said I, "but I never thought more of it than I should of Tosspot's crockery cart, or Warren's matchless blacking van."
' "I could let you in for a good thing", observed the stranger musingly.
' "Haste me to know it, that I with wings as swift as meditation or the thoughts of love, may—*jump at it*", exclaimed I.
' "I find thee apt", rejoined the stranger, rising and extending his right arm, saying—

> ' "*And duller should'st thou be than the fat weed
> That roots itself in ease on Lethe's wharf,
> Would'st thou not stir in this.*" '

' "Oh, my prophetic soul! my uncle!" exclaimed I, interrupting him; "if it wasn't for that black pow and those d—d heavy brows, I'd swear you were my old friend Jack Rafferty, late of the Adelphi Theatre."
' "You have me!" said he, pulling off the wig and appurts with one hand and grasping my hand with the other. Sure enough, there stood old bald-headed Jack, with his little ferrety eyes peering at me with the great black brows still above them. Having taken these off and put them carefully in his pocket-book, he again shook hands, and asking for a squeeze of the old comforter, we stirred the fire, put on the kettle, and prepared for hot stopping.
' "Bill", said he, as soon as he had got the brew to his liking, and one of my best Woodvilles in his mouth, "one good turn deserves another."
' "Undoubtedly", said I, "as the tailor observed when he turned the old trousers a second time."
' "Ah!" said he, "you're just the same old cove that ever you were. How are you off for blunt?"
' "D—d badly", said I; "should be glad to join you in raising a mortgage on our joint industry."
' "Well, never mind", said he, chuckling, "you did me a good turn when that wicked bailiff, Levy Solomons, came to

take me for the butter bill, and I haven't forgotten it. By Jove! I fancy I hear him blobbing into the rain-water tub at this moment. I've seen queer days since then", added he thoughtfully; "been all through the Disunited States, Canada, Columbia, and I don't know where, shipwrecked twice, gaoled thrice, tarred and feathered besides. Hard life a player's, forced to appear merry when we're fit to cry; however, that's all done—I've turned over a new leaf—I'm in the respectable line now, and hearing that your occupation in Lincoln's Inn's gone, why I've just stepped in, as Paul Pry would say, to see if I could do anything for you in the respectable line too. You see", said he, "the way for talented men like us to prosper is to take the folly of the day and work it. I saw this in the nigger times. Lord, if the compensation money had been taken direct from the pockets of the people, instead of passing through the filtering bag of Parliament, it would have been a good workable subject to this day. John Bull is a great jackass—a thick-headed fool. Unless you empty his breeches pocket before his face, and say, 'Now, John, I take this shilling for the window tax, this for the dog tax, this for the gig tax, and this for the nigger tax', you can't make the great muddle-headed beast believe he pays anything for the nigger tax, and so by making it a parliamentary grant, opposition was lost, and with it as fine a field for enterprise as ever was seen. However, it's no use crying for spilt milk. Go ahead's my motto, as they say in the Disunited States. But to business.

'"The new light is the Corn Laws. There's more sense in this than there was in the nigger question, because if you can persuade a man he'll get a fourpenny loaf for twopence, you show him something to benefit himself, which you couldn't do in the case of the great Bull niggers, that he had never seen or cared to set eyes on. Still John shows his stubbornness, and hangs back as if he thought the Repealers were the only people that would get the fourpenny loaf for twopence. It is to rouse the animal, and convince him that for once there is such a thing as pure disinterestedness in the world, that the League is bestirring itself; and now, my old friend Bill", continued he, "for the service you did me, by popping the bailiff over the head in the tub, I've come to offer to recommend you, as a man of very great talent, eloquence, experience, and I don't know what; in fact, to supply the vacuum there must necessarily be in the heads of men who are fools enough to subscribe their money to force a benefit on people that they don't want.

'"The League is about to enlighten the country—north, south, east, and west—from the Orkneys to Portsmouth, from Solway Firth to Flamborough Head—all are to be visited by men of mettle like ourselves, and if we don't astonish the natives, why my name is not Jack Rafferty."

'"Faith", said I, "Jack, I'm not nasty particular, and never was about making money, especially at the present time, for to tell you the truth, I'm as near in Short's Gardens as ever I was in my life; but the devil and all is, I know nothing about either corn or the corn laws, and hardly know wheat when I see it."

'"That's nothing", said Jack; "you've a quick apprehension and a ready tongue—lots of jaw—and *that's* what the League want. You'll have plenty of time to study your part, and rehearsals over and over again. Zounds, man, it's the easiest thing in life! Instead of appearing in one character on Monday, another on Tuesday, a third on Wednesday, a fourth on Friday, and a fifth on Saturday, and having to study and cram and rehearse for them all, here you have nothing to do but repeat the same old story over and over again, which comes as pat off the lips as a child's church catechism. 'Infamous aristocracy' — 'iniquitous' — 'ruinous starvation' — 'landlord-supporting tax' — 'blasted Quarterly' — and all that sort of thing. Whatever is wrong, lay it to the corn tax. If a man can't pay his Christmas bills, attribute it to the bread tax; say the landlords have grabbed a third of his income. Tell the ship-owners their interest is ruined by the monopolists—nay, you may even try it on with the farmers, and say you verily believe they would be benefited by the abolition of the corn laws; that you really think our climate and system so superior, that they would drive foreign grain out of the market, just as our fat Durhams and Devonshires beat Sir Robert's Tariff fat cattle out of the shambles. In fact, you may say almost anything you like; and should anyone oppose you, you will always be ready with a cut and dried answer, which, with an easy delivery, will put your cleverest unprepared arguer quite in the background."

'Just then, in came Mrs. B. "Cleopatra, my dear, here's our old friend, Rafferty", said I.

' "What, Jack!" exclaimed she, "that robbed the treasury at the Adelphi?"

' "Hush!" cried I. "Jack's respectable. *Encore* the brandy."

'Well, the upshot of it was, that the next day I attended a meeting of the League at the British Hotel, in the best apparel I could muster—light blue, buff vest, drab tights, best Hessians, tartan cravat. Joey Hume was in the chair, and as soon as ever I saw that, I determined to be stiff.

' "Who have you got there, Mr. St. Julien Sinclair?" (for that is the name Jack goes by)—asked Joe, as we advanced to the table.

' "Mr. William Bowker", replied he.

' "The same of whom you spoke at our last meeting?" inquired Joe.

' "The same", answered Mr. St. Julien Sinclair.

'Jack had primed me pretty well on the road what I should say, in case they examined me; but I suppose, being well recommended, or knowing it must come to that at last, they thought it better to dispense with all humbug, and having ascertained that I was perfectly disengaged, and ready to embark in the cause, they said that the Council of the League had determined to sectionize the kingdom; to enlighten the lower orders on the monstrous iniquity of the bread tax, and the great advantages of a free trade in corn. That they had been at it for some time without producing much effect, but they had now got a new dodge which they thought would tell. This was, that instead of single-handed lecturers, like Jack Rafferty, going about doing as they liked, and reporting what they pleased, that the leaders of the League should take the thing in hand, distribute themselves over the land along with ladies and lecturers, and make a regular crusade against the monopolists. Lecturers, it seems, they had not much difficulty in getting, indeed I should wonder if they had, for eight guineas a week and one's travelling expenses are not picked up every day—but the ladies there had been some trouble about. However, as they thought they could not dispense with the influence of the fair sex, they have accommodated matters by hiring a certain number of females who are to take superior characters, just as Jack Rafferty took the part of Mr. St. Julien Sinclair. To each lecturer, therefore, there is to be attached a leader and a lady; and the company are building a lot of Whitechapels, capable of carrying three with their luggage, and we are to be allowed ten shillings a day for a horse to pull them about. There will be suitable devices, with mottoes, such as "DOWN WITH THE BREAD TAX!" — "FOOD FOR THE MILLION", &c., &c., along the sides of the vehicles, which are to be painted sky-blue, with red wheels, picked out with green. They will be labelled behind in statuteable letters—

' "GREAT NATIONAL ANTI-CORN-LAW ENLIGHTENMENT CART"
' "FORMS FOR PETITIONS SUPPLIED!"

'I think that is all I've got to say, except that I hope your new purchase is to your liking, and that Mrs. Jorrocks approves of the house as much as she did of her mother's at Tooting. Should there be anything I can do for you in town, pray let me know; and after I leave, Cleopatra or Susan will be glad to do their best for either Mrs. Jorrocks or you, to whom we all beg to present our most respectful compliments, and I have the honour to subscribe myself, Dear sir, your humble and obedient servant,

'Wm. BOWKER, L.G.A.C.L.L.A.
'*Lecturer to the Grand Anti-Corn-Law League Association.*'

(*Hillingdon Hall*, 1845.)

A LITTLE POLITICAL ACCOMMODATION

*I*T was turning dusk as Mr. Smoothington reached the hill above Sellborough on his way back from Donkeyton Castle, but the wind setting towards him, sounds of music and drunken revelry were borne on its wings.

Mr. Bowker had made a grand entry into the town at three o'clock, amid the most enthusiastic demonstrations from the populace. They met his carriage at the turnpike gate, on what had been the London, but was now called the Smoke Station road, and, having taken the four panting posters from it, had drawn him through all the principal streets, preceded by numerous splendid banners, and two bands of music.

The honourable gentleman had made a most favourable impression. He was dressed in the height of fashion—a mulberry-coloured frock-coat with a rolling velvet collar, and a velvet waistcoat of a few shades brighter colour than the coat; an extensive flowered satin cravat, with massive electrotype chained pins, fawn-coloured leathers, and Hessian boots. His touring excursions having supplied him with an abundant stock of health, he presented a very different appearance to what the generality of country people imagine a London merchant to be like.

Altogether, he created an indescribable sensation; and as he passed along, standing up in his barouche, bowing gracefully to the ladies, they waved their handkerchiefs, and declared he was 'a most charming man'. Then, when he got to the 'Duke's Head', he appeared in the balcony of the drawing-room, and addressed them on the importance of the privilege they would soon be called upon to exercise. After alluding touchingly to the lamented death of Mr. Guzzlegoose, he called upon them to exercise the elective franchise in such a way as would be beneficial to themselves, their posterity, and their country at large, when the elegance of his manner, and the graceful flourishes of his lavender-colour kidded hand, carried all before it, and men, women, and children hurrahed, and shouted 'Bowker for ever!'

But when he came to expatiate on their wrongs, pointed out the injury they sustained by the operation of the Corn Laws, exposed their exclusive workings for the benefit of the landlords, and called upon them to support a candidate favourable to their immediate and total repeal, the enthusiasm of the mob knew no bounds, and every hand was held up in favour of Mr. Bowker—'Big-loaf Bowker', as he christened himself.

After partaking of some light refreshment, he then commenced his canvass, amid the ringing of bells, the rolling of drums, the twanging of horns, and the shouts of the populace; and if unregistered promises could have brought him in, Mr. Bowker would certainly have been member for the county.

Thus he spent the day—shaking hands—praising and admiring the children, chucking damsels under the chin—promising all things to all men. At length, tired of the din and flurry of the proceedings, Mr. Bowker was glad when five o'clock came; and with his old friends Mr. St. Julien Sinclair, and his committee, Mr. Lishman, a bankrupt baker, Mr. Grace, an insolvent painter, Mr. Moss, a radical schoolmaster, and Mr. Noble, a sold-off farmer, he left the streets to enjoy the evening repast at the 'Duke's Head'. The landlord, Mr. Tucker, in a white waistcoat, followed by his waiter and boots in their best apparel, met the distinguished guests at the door, and conducted them to the drawing-room.

Mr. Bowker, after begging to be excused a few minutes while he went and washed his hands (a thing his committee never thought of doing), retired to his bedroom, and made a perfect revision of his costume. When he returned he was in an evening dress, smart blue coat with club buttons and velvet collar and cuffs, white neckcloth, superbly embroidered waistcoat, with black silk tights, and buckled shoes. He dangled a pair of primrose-coloured kid gloves in his hand.

'We may as well ring for dinner', observed the florid swell, entering the drawing-room, and surveying the seedy crew sitting round. He gave a pull that sounded through the house.

The dinner was quickly served, and as quickly despatched by the hungry guests, several of whom had not tasted meat for a week. Champagne, hock, claret, sparkled on the board, and was swallowed by some whose stomachs were much more accustomed to beer.

As evening shades made the sherry indistinguishable from the port or claret, and Mr. Tucker, in obedience to the Squire of Whetstone Park's summons, was bearing a branching candelabra through the passage on his way upstairs, Mr. Smoothington arrived at the door of the hotel, and begged Mr. Tucker to carry his card up to Mr. Bowker.

Accordingly that functionary did so.

'Smoothington!' said Bill, glancing at the gilt-edged pasteboard with the easy indifference of a man accustomed to callers. 'Smoothington! who is he?'

'Smoothington!' exclaimed the bankrupt baker and sold-off farmer, each of whom were undergoing Mr. Smoothington's polite attentions.

'Is he an elector?' inquired Bill, considering whether he should see him.

'He's the Duke of Donkeyton's solicitor', replied mine host.

'Indeed!' observed Mr. Bowker; adding, 'show him into a room, and I'll ring and let you know when it's convenient for me to see him.'

'Yes, sir', said Mr. Tucker.

'Help yourselves, gentlemen', said Mr. Bowker, filling his glass and passing the bottle.

'We'd better cut our sticks, I think', observed the baker, significantly, to the Corn-Law-ruined farmer.

'I think so too', replied the latter.

'And I'll go with you', added Mr. Grace, the insolvent painter, who lived in a house belonging to the Duke.

'Oh no, gentlemen', said Bill, 'don't disturb yourselves—don't disturb yourselves—I'll receive Mr. Smoothington in the other room.'

'We'll go there!' exclaimed all three—'we'll go there!' thinking to avoid meeting Mr. Smoothington on the stairs.

'Take a bottle of wine with you!' said Bill, pushing the port towards them.

'Thank ye—we'd prefer glasses and pipes', observed Mr. Lishman.

'Ah, you are the right sort, I see', replied Bill; 'nothing like baccy.'

They all then bundled out.

'Just put the table right, and take these dirty plates away', said Mr. Bowker, as the landlord answered the expected summons.

'Now, give a couple of clean glasses, and tell Mr. Smoothington I shall be happy to see him', said Bill, twirling the card about.

Mr. Smoothington's creaking boots presently sounded on the stairs as he ascended two steps at a time. Another moment, and he was bowing and scraping in the room.

'Mr. Smoothington, I believe', said Mr. Bowker, rising and bowing to the stranger.

'The same', replied the man of law, making one of his best Donkeyton Castle bows, and laying his hand on his heart.

'Pray, be seated', said Mr. Bowker; 'pray, be seated', said he, laying his hand on the back of the chair, by the clean glasses and plate.

Mr. Smoothington put his hat under the chair, and obeyed the injunction.

'Take a glass of wine', said Mr. Bowker, passing the bottle across. 'That's the claret without the label; you'll find it better than the port.'

'Thank you, sir', said Mr. Smoothington, helping himself to the claret.

'Confound these country inns', observed Mr. Bowker, 'they've no notion of doing things properly. Only fancy! They've sent up champagne without being iced!'

'Indeed!' exclaimed Mr. Smoothington.

'Did, 'pon honour', said Bill, with a shake of the head. 'The claret's not what it should be, but the landlord says it's the best he can give. I'm sorry I can offer you no better dessert than these filberts and biscuits', added he; 'but to tell you the truth, I've had the misfortune to lose my footman and part of my luggage.'

'Indeed!' exclaimed Mr. Smoothington, with a look of concern.

'He's either left behind at a station, or carried past the right one; at all events, when I wanted him he was not to be found. The worst of it is', added Bill, 'he had a couple of pine-apples and some fine grapes, that my gardener—poor fellow—thought would be a treat for me in the country.'

'Indeed!' rejoined Mr. Smoothington; 'that *is* a loss'; as much as to say, the footman was nothing.

'Why, it is a loss, as things stand', said Bill, 'for I should have liked to have offered you a slice. As for myself, I care nothing about them; but we are supposed to grow the finest in England.'

'You are very kind, I'm sure', replied Mr. Smoothington; adding 'have you much glass?'

'Three houses, I think', said Bill; 'three pineries—that's to say three vineries; peach-house or two. But I care very little about a garden.'

'Pay more attention to your park, perhaps', observed Mr. Smoothington.

'Ay, *there you have it*!' said Bill, brightening up; 'there you have it', repeated he. 'My friend, Lord Scampington, pays me the compliment of saying I've the finest venison in England.'

'Have you indeed?' exclaimed Mr. Smoothington, who dearly loved the cut of a haunch, particularly when he could get a glass of Burgundy after it.

'Help yourself', said Mr. Bowker, pushing the bottles towards him, thinking his friend would want something to wash the lies he was telling him down with. Mr. Smoothington did as desired. Pending the gulp which followed, he bethought him of business.

'I hope you are not tired with the exertion of your canvass', observed Mr. Smoothington, rubbing hand over hand.

'Why, not tired', said Bill, with an air of indifference; 'not tired—*rather* bored.'

'You are on the Repeal interest, I perceive', said Mr. Smoothington.

'Repeal decidedly', replied Bill. 'By the way, did you see my little English and big American loaf dangling from the balcony as you came in?'

'It was dusk', replied Mr. Smoothington; 'and there was a great crowd about.'

'Looking at it, I dare say', said Bill. 'The best dodge yet.'

'The Corn Laws must be repealed', observed Mr. Smoothington; 'every thinking man must be satisfied of that. I think, however, it is rather a pity for two champions to start in the same cause when only one can come in.'

'How so!' exclaimed Mr. Bowker; adding, 'what! is there another Richard in the field?'

'The Marquis of Bray and yourself', observed Mr. Smoothington.

'The Marquis of Bray's the other way', replied Mr. Bowker.

'Pardon me', rejoined Mr. Smoothington.

'He wouldn't declare himself, at all events', observed Bill, 'and we politicians generally consider those who are not for us are against us.'

'It was partly out of delicacy to the memory of Mr. Guzzlegoose, and partly a mistake of mine!' observed Mr. Smoothington.

'How so?' asked Bill, filling himself a bumper, and passing the bottle.

'Why, I prepared his Lordship's address, the draft of which I now produce', said Mr. Smoothington, diving into the back pocket of his coat, and producing some ominous red-taped papers. 'In this draft as you will perceive', continued he, opening it out, 'distinct allusion is made to *all restrictions* on trade, including, of course, the Corn Laws; but, by an unfortunate clerical error, the important sentence was omitted, and the bill printed and posted without—'

'That's very odd', observed Mr. Bowker; adding, 'shows great inattention on—'

'I was called away at the moment to attend to a relation who was dying', interrupted Mr. Smoothington.

'Well, but why didn't the Marquis answer the League letters'? asked Bill, adding, 'great body of that sort is entitled to respect, even from a Marquis.'

'That was a pity, certainly', replied Mr. Smoothington. 'If I had been at home it would have been otherwise. These young men, you see, are unused to business—inattentive. I can answer for it, however, that not the slightest disrespect was meant to the League."

'*Hum!*'considered Bill.

'It certainly seems a pity', continued Mr. Smoothington, 'that two candidates of the same opinions should offer themselves for the same seat; to say nothing of the probability, nay, certainty, of the Tories putting up a man, and getting it from them.'

'I'm not afraid of the Tories', replied Bill, 'as a party they are contemptible against the League.'

'Single-handed, they are, I dare say', agreed Mr. Smoothington, 'but if the League interest is split, a very small party will defeat it.'

'True!' observed Mr. Bowker, seeing how the thing would cut. 'Well, then, the best thing would be for the Marquis of Bray to retire', added he; 'can be no difficulty about that, you know.'

'Except that the Marquis's interest has always been paramount in the county.'

'Time there was a change then', observed Bill. 'The Reform Bill ought to have all that put right.'

'I'm afraid I could hardly advise the Marquis to retire', observed Mr. Smoothington after a long pause.

'You can hardly expect me to do it, I think, after all the expense I've incurred', replied Mr. Bowker.

'Perhaps we could accommodate matters', suggested Mr. Smoothington, helping himself to the proffered bottles. 'The Duke has a great interest in the neighbouring borough of Swillington, and a dissolution can't be far off; his interest there might return you comfortably for a long session, without trouble or expense.'

Mr. Bowker sat silent, apparently considering the matter.

'County representations are very troublesome', observed Mr. Smoothington; 'people never done asking—schools, churches, hospitals, infirmaries, races, plays, farces, devilments of all sorts—no gratitude either. At Swillington there's nothing but a dinner, and a guinea a-head to the voters; five hundred pounds would do it.'

'I should still lose all the expenses I have been at here', observed Mr. Bowker.

'That could be accommodated too', replied Mr. Smoothington.

'Consider the trouble, though', bristled Mr. Bowker. 'What can compensate me for my trouble, mental anxiety, and so on?'

'True!' assented Mr. Smoothington, unable to price it.

'Separation from family', urged Mr. Bowker.

'Very true', replied Mr. Smoothington.

'Leaving one's own comfortable home for a filthy frowsy inn, where they haven't even common decency, I may almost say, necessity of life, ice for champagne.'

'This, I fear is beyond the reach of our control', observed Mr. Smoothington, rolling his hands over and over.

'Money can't put that right', said Mr. Bowker.

Mr. Smoothington shook his head. 'Its an unfortunate thing that the Marquis and you should have come in collision', said he.

'It is', said Mr. Bowker, '*most* unfortunate.'

'The Duke is a most amiable person', observed Mr. Smoothington; so is the Duchess; you'd like them if you knew them.'

'Faith, I'm not a great man for the nobility', observed Mr. Bowker. 'Am very much of an old friend of mine's way of thinking; who says that they first tried to make towels, and then dish-clouts of one.'

'The Duke of Donkeyton doesn't', replied Mr. Smoothington; 'he's always the same.'

'Good fellow, is he?' asked Mr. Bowker.

'*Very*', replied Mr. Smoothington.

'And the Marquis, what's he like?' asked Mr. Bowker.

'Very fine young man'; said Mr. Smoothington.

'Indeed!' mused Mr. Bowker.

'Perhaps you'd go over with me and talk to the Marquis?' observed Mr. Smoothington, after a pause.

'Why, I don't know', replied Mr. Bowker; 'I dare say we can do all he could.'

'No doubt', rejoined Mr. Smoothington; 'no doubt. The Duke will ratify whatever I do.'

'You are his factotum, I suppose', observed Mr. Bowker.

'The Duke does nothing without consulting me', replied Mr. Smoothington, with a self-complacent smile.

'It's an awkward business', mused Mr. Bowker; 'commenced my canvass — extremely popular — great disappointment — enormous expense.'

'The expense should be *no* object', replied Mr. Smoothington, 'if only you could get over the rest.'

Mr. Bowker meditated.

'Nay, I don't want to drive a hard bargain', at length said he, with an air of indifference.

'It's only *right* you should not be out of pocket', replied Mr. Smoothington; 'indeed, I should consider it my duty to see that you were not, the mistake having originated partly with myself.'

'Well', said Mr. Bowker, again helping himself, and passing the bottle, 'your proposition appears reasonable—fair, I may say.'

'I am glad you think so', replied Mr. Smoothington; 'there is only one way of dealing with gentlemen like you.'

'Let me see', said Mr. Bowker, rubbing his hands; 'it is that the Duke returns me for Swillington at the General Election, and pays my present expenses—that's to say, up to to-night?'

'I'll agree to that on behalf of his Grace', replied Mr. Smoothington, bowing and helping himself.

'It may save trouble', said Bill, 'if I take a sum down. There are expenses in town as well as here', added he.

'As you please', replied Mr. Smoothington. 'What shall we say?'

'Put it in at your own figure', said Bill, with a shrug of his shoulders, and an air of indifference. 'A thousand! *say* a thousand!' added he.

This was a good deal more than Mr. Smoothington expected; but coming from a man with three pineries, and the best venison going, he thought it better to close than to haggle; especially as he was dealing for a Duke.

'Agreed', said Mr. Smoothington.

'Help yourself', said Mr. Bowker, again passing the bottle, 'and drink success to the Marquis of Bray.' Mr. Bowker drank it in a bumper.

'His lordship will be much flattered when I tell him the

compliment you've paid him', said Mr. Smoothington, filling his glass and doing the same.

'You may as well give me a cheque for the money to-night', said Bill, 'and let me get out of this noisy place before they resume their racket in the morning.'

'With all my heart', replied Mr. Smoothington, thinking he had better clench the bargain and get an agreement of resignation at the same time. Pens, ink, and paper being then produced, Mr. Smoothington filled up a cheque for the required sum, and took a memorandum of the agreement from Mr. Bowker, who got a duplicate signed by Mr. S., on behalf of the Duke of Donkeyton.

Exulting in his diplomacy, Mr. Smoothington shortly after backed out of the room, not, however, without receiving a pressing invitation from Bill to visit him at Whetstone Park.

With a somewhat swimming head, Mr. Smoothington descended the inn stairs; and, after ordering an express to come to his house, as soon as he could get ready, he sat down at his desk at home to write his letter to Donkeyton Castle just as the market-place clock chimed midnight.

(*Hillingdon Hall*, 1845.)

GRASS-LAND AND ARABLE

THERE is a regular rolling stock of bad farmers in every country, who pass from district to district, exercising their ingenuity in extracting whatever little good their predecessors have left in the land. These men are the steady, determined enemies to grass. Their great delight is to get leave to plough out an old pasture-field under pretence of laying it down better. There won't be a grass-field on a farm but what they will take some exception to, and ask leave to have 'out' as they call it. Then if they get leave, they take care never to have a good take of seeds, and so plough on and plough on, promising to lay it down better after each fresh attempt, just as a thimble-rigger urges his dupe to go on and go on, and try his luck once more, until land and dupe are both fairly exhausted. The tenant then marches, and the thimble-rigger decamps, each in search of fresh fields and flats new.

Considering that all writers on agriculture agree that grass-land pays double, if not treble, what arable land does, and that one is so much more beautiful to the eye than the other, to say nothing of pleasanter to ride over, we often wonder that landlords have not turned their attention more to the increase and encouragement of grass-land on their estates than they have done.

To be sure they have always had the difficulty to contend with we have named, *viz.*, a constant hankering on the part of even some good tenants to plough it out. A poor grass-field, like Gay's hare, seems to have no friends. Each man proposes to improve it by ploughing it out, forgetful of the fact, that it may also be improved by manuring the surface. The quantity of arable land on a farm is what puts landlords so much in the power of bad farmers. If farms consisted of three parts grass and one part plough, instead of three parts plough, and one part grass, no landlord need ever put up with an indifferent, incompetent tenant; for the grass would carry him through, and he could either let the farm off, field by field, to butchers and graziers, or pasture it himself, or hay it if he liked. Nothing pays better than hay. A very small capital would then suffice for the arable land; and there being, as we said before, a rolling stock of scratching land-starvers always on the look-out for out-of-order farms, so every landowner should have a rolling stock of horses and farm-implements ready to turn upon any one that is not getting justice done it. There is no fear of gentlemen being overloaded with land; for the old saying, 'It's a good thing to follow the laird', will always insure plenty of applicants for any farm a landlord is leaving—supposing, of course, that he has been doing it justice himself, which we must say landlords always do; the first result we see of a gentleman farming being the increase of the size of his stock-yard, and this oftentimes in the face of a diminished acreage under the plough.

Then see what a saving there is in grass-farming compared to tillage husbandry; no ploughs, no harrows, no horses, no lazy leg-dragging clowns, who require constant watching; the cattle will feed whether master is at home or polishing St. James's Street in paper boots and a tight bearing-rein.

Again, the independence of the grass-farmer is so great. When the wind howls and rain beats, and the torrents roar, and John Flail lies quaking in bed, fearing for his corn, then old Tom Nebuchadnezzar turns quietly over on his side like the Irish jontleman who, when told the house was on fire, replied, 'Arrah, by Jasus, I'm only a lodger!' and says, 'Ord rot it, let it rain; it'll do me no harm! I'm only a grass-grower!'

(*Ask Mamma*, 1858.)

ENTER THE RAILWAY TRAIN, AND THE LONDON CLUB

TIME was—before the establishment of railways—that the Squires used to respond to the call of their chiefs with the greatest alacrity, but the whistle of the engine had somewhat dispelled the authority of the leaders, and made men think more for themselves than they did. In truth, there is perhaps no class of Her Majesty's subjects more benefited by the introduction of railways than the country gentlemen generally, who too often, after what used to be called the 'Grand Tour', buried themselves and their usually good educations in remote country places, there to marry 'neighbours' bairns', and perpetuate the practice. Now they fly about the world, here and there and everywhere, importing ladies from all parts, making the whole kingdom but as one country, while the lists of members of the various Clubs show that they are not indifferent to the attractions of the capital. The very thing has come to pass that was predicted when stage-coaches were first established some two hundred years ago, namely, that 'country gentlemen and their wives would get easily and cheaply conveyed to London', without the remainder of the prophecy, however, being fulfilled, namely, 'that they would not settle quietly at their homes in the country afterwards', for whole families whisk about in all directions, and feel all the better for the change, enjoying their spacious homes the more from having perhaps put up with contracted quarters elsewhere.

Heaven help the parties' idea of ease who attributed anything of the sort to even the latest and best of the old stage-coaches, let alone the ponderous, unwieldy vehicles that first ploughed the bottomless roads, turning up the great boulder-stones like flitches of bacon, and taking the liberal allowance of from twelve to sixteen days in performing the journey between London and Edinburgh! Dr. Johnson, we make no doubt, described very accurately what they were in his time, when he boasted that he had travelled from London to Salisbury in a day by the common stage, 'hung high and rough'. The Doctor's observation, that a postchaise had jolted many an intimacy to death, was doubtless very correct also. Who hasn't a lively recollection of the musty old horrors? Talking of travelling, there is, or was, a notice in the coffee-room of the Black Swan Hotel at York, stating that a four days' stage-coach would begin to run (crawl, would perhaps have been a more proper expression), on Friday the 12th of April, 1706.

'All that are desirous to pass from London to York', continues the advertisement, 'or from York to London, or any other place on that road, let them repair to the Black Swan in Holborn, in London, or to the Black Swan in Coney Street, in York.

'At both which places they may be received in a stage-coach every Monday, Wednesday, and Friday, which performs the whole journey in four days (if God permits). And sets forth at five in the morning. And returns from York to Stamford in two days, and from Stamford by Huntingdon to London in two days more. And the like stages on their return. Allowing each passenger 14 lbs. weight, and all above 3d. a-pound.'

Rather a diminutive allowance for a modern exquisite's luggage.

As time advanced, the pace certainly improved, but even up to the last of the coaches, they were five times as long as the rail.

In truth, the country gentlemen were a land-locked, leg-tied tribe, before the introduction of railways—coaching was uncomfortable, and posting expensive, besides which a journey took such a time. There was no running up to town for a week in those days. It took the best part of a week coming from a remote country to make the journey, and recover from the effects of it. No wonder the gentry did not make them very often, and contented themselves with their country towns instead of the capital. They were somebody in them, but nobody when they got into London. It seems rather strange, though, that even in those days, when transit was so slow and expensive, and men had to live so long on the road, that there were always plenty of country gentlemen ready to contest their respective counties, though the cost was frightful, and the poll as lingering as the coaches.

Now, when both counties and costs are curtailed, and transit so quick, there is great difficulty in getting country vacancies filled by resident gentry as they occur. The fact is, the world is so opened out, that every man who has a taste for travel, or who can sit a horse, or walk a moor, thinks he can employ his time and money better than in paying for working for other people. Members ought to be elected free of expense, and then let them work for nothing if they liked. It is singular that some of the greatest screws—some of the most determined 'nothing for nothing', and uttermost-farthing men, are now the greatest spendthrifts in the matter of electioneering expenses. And the humiliating part of the matter is, that men who question and fight every farthing in their respective trade transactions, will part with thousands upon thousands in the wildest, blindest way, and declare that they hadn't the slightest idea the money was for anything but legitimate expenses! Why didn't they see to its application then?

But to return to the Old Squires. Another safety-valve that the gentry of the old school had for emitting the steam of their wealth, besides keeping hounds and electioneering, was in huge house-building—they built against each other. If Squire Fatfield built a great staring house, Squire Flaggon would follow suit with a bigger, and Squire Jollybuck would cap Squire Flaggon with a larger still. Now building a big house, and buying a big house, are two distinct things; for the builder of a big house is expected to live in it, and maintain a suitable establishment, while the buyer of a big house can shut up as much of it as he finds is too large for his purpose. Then the larders, and the cellars, were expected to correspond with the houses, the characters of the owners depending a good deal on the strength of their taps, while the conviviality of the dining-room always found a hearty response in the servants' hall—masters and butlers considering it a reproach to let anyone leave the house sober. These hospitalities expired together, French wines superseding the glorious old port, and railways opening out other means of expenditure than upon malt liquor for gratuitous distribution. A country house in former days was little better than a great unlicensed inn—everything was taken in that arrived, and everybody had to be refreshed that came. We have heard of a gentleman—not an M.P., or a man of large fortune either—whose brewer's bill for a single year amounted to no less a sum than eight hundred pounds!

In thus noting the manners and customs of a bye-gone day, we must not omit to do justice to the merits of the port wine, which certainly was excellent. There was no buying of two-dozen hampers in those days; every man had his stock of port wine in wood as well as in bottle, and that in the wood was not advanced to the bottle before a long probationary process. Being at length bottled, it would lay many years in its bin mellowing for use, an occasional bottle being produced to competent judges to see how it was advancing, and then when it was at length pronounced 'fit', it was 'drunk on the premises' without further to-do. Port was the staple beverage in those days, fine clear ruby-coloured wine, not a headache in a hogshead of it as the old ones used to say, and certainly they tried it at high pressure. They generally drank out of small glasses, so small indeed as to be insignificant, and a man helped himself almost incontinently, as the oft-recurring bottle passed round. In the midst of mirth and conversation, one man is very apt to do what another does, and it is not till the next morning that he becomes sensible of any excess.

There was no blowing men out with Champagne or sparkling Moselle during dinner then, as there is now; Sherry and Madeira were the regulation wines, varied perhaps latterly with a little of what the Yorkshire farmer called 'Bluecellas'; but the dinner wines were rarely taken into account, the night's consumption being calculated solely on the port. In fact, the real business of the evening did not commence until the ladies (or as they say in Courts of Justice, women and children) had withdrawn from the dining-room; then the horse-shoe table would be brought out, the fire stirred up, the log put on, and everything arranged for a symposium.

We can fancy the surprise and indignation of a party of these worthies at the intrusion of the three-quarters drunken butler, and the half-drunken footman, with coffee, at the end of half an hour after they had got so settled. We think they would go out faster than they came in. But we will not imagine anything so monstrous and inhospitable. No; the party sit true to their glasses, the bottle circulates briskly, the glasses are fairly filled to the brim, and as fairly drained, and a couple of hours glide away, amidst jokes, songs, and sentiments, ere there is even a summons from the ladies. If the jokes were not very new, they answered just as good a purpose as if they were, and it shows a kindly disposition to greet an old friend with a laugh. There was no *Punch* in those days to supply the weekly stock of fun, and the papers were small, and deficient of news. No family breakfast table-cloth-like sheets, with information from all parts of the globe. But if the size was small, the price was large; sevenpence being charged, some forty years ago, for a four-columned London paper of four pages. A quick reader would skim through one of them in five minutes, for the type was bold and well-leaded. The country papers were worse, and contained little but advertisements: 'Horse stolen', 'Hay for Sale', 'Green Dragon Inn to Let', 'Main of Cocks to be Fought', 'Gout and Bilious Pills', 'Cornhill Lottery Tickets', 'Fire and Life Assurance Offices', all well spread out in the most liberal, amplified way; mixed with murders, inquests, and a very slight sprinkling of political and parliamentary news. No wonder that people were thrown on their own and each other's resources for information and amusement.

Now, every pursuit and calling has its organ, all admirably conducted, and published at very low prices, so that a modern squire can select such papers as suit his taste, and have his non-eating, non-drinking guests down by each post, whom he can lay aside when he's had enough of their company, which he can't do by a tiresome chattering guest, who can neither talk nor hold his tongue. Some squires are not very lively.

The establishment of the Penny Post, and the liberal scattering of post-offices too, has been a wonderful boon to country gentlemen, indeed to all sorts and conditions of people; but the old squires being about the only people in the country who received letters, or who, perhaps, could read them when got, were often sadly put to in the sending long distances for them. To be sure many of them did not care much about getting them, and there are even some now, who if they happen to leave home for a few days, won't have them forwarded on to where they are.

The grand, the crowning benefit of all however, were railways. Without them, cheap postage, cheap papers, cheap literature, extended post-offices, would have been inefficient, for the old coaches would never have carried the quantity of matter modern times has evoked. Who does not remember the last spasmodic efforts of the unwashed, worn-out old vehicles, and weak horses to compete with the accumulating traffic in the neighbourhood of a newly-making line—amid the anathemas of coachmen and guards, and their brandified predictions of a speedy return to the road? But at a certain hour on a certain day, without noise, or boast, or effort, came the smoothly-gliding engine, whisking as many passengers along as would have filled the old coaches for a week, unlocking the country for miles, and bringing parties within a few hours of each other who had formerly been separated by days. Large, roomy, prebendal stall-fitted-up-like vehicles, usurped the place of little stuffy, straw-bedded stages, into which people packed on the mutual accommodation principle, you letting me put my arm here, I letting you put your leg there. So they toiled on through a live-long day, cramped, squeezed, and confined, making about the same progress that they do now in a couple of hours with the greatest ease and enjoyment. Independently of the saving of time, railways may be looked upon as downright promoters of longevity, for assuredly a man can do and see twice as much as he formerly could without; so if Squire Mistletoe lives to seventy or eighty, he will be entitled to have put on his monument that he died at a hundred and forty, or a hundred and sixty, as the case may be. Squire Mistletoe can run up to town fifty times for once that his father did, and feel all the better instead of all the worse for the trip.

The next greatest boon to railways that modern squires have to be thankful for, is the great multiplication of London Clubs.

Without Clubs, the railway system would have been incomplete. After such luxurious travelling a man requires something better than the old coaching-houses—the Bull-and-Mouth, the Golden Cross, or even than the once prized Piazza, with its large cabbage-smelling coffee-room. A night at the old Bull-and-Mouth, with its open corridors, was a thing not to be forgotten. The railway companies, to be sure, anticipated the want, and built spacious hotels at their respective termini, the Piazza became a Crystal Palace, and the Bull-and-Mouth changed its ugly name! but disguise it as you will, an hotel is an hotel, and an Englishman cannot make himself believe that it is his home.

Then these railway houses are all out of the way of where pleasure-seeking people want to be, and though a party's requirements are fairly supplied, yet these hotels hold out no inducement for a run up to town for the mere pleasure of the thing. This is what the Clubs do. They invite visits. A man feels that he has a real substantial home—a home containing every imaginable luxury, without the trouble of management or forethought—a home that goes on as steadily in his absence as during his presence, to which he has not even the trouble of writing a note to say he is coming, to find everything as comfortable as he left it.

No preparation, no effort, no lamps expiring from want of work; good fires always going, good servants always in attendance, everything anticipated to his hand. Verily, a member of a Club may well ask, 'What are taxes?'

Clubs in fact, are the greatest and cheapest luxuries of modern times. We have before us the balance-sheet of one of the largest Clubs in London, whose income is some fifteen thousand a-year, which of course is all spent inside the house, there being no carriages, no horses, no coachmen, no grooms, no valets; nothing but butlers, waiters, cooks, housemaids, what are generally called menial servants, in fact. Of this £15,000, salaries and wages come to between £2,000 and £3,000 a-year; lighting, £1,000; fuel, £500; liveries, £400; washing nearly as much; and for some eight or ten pounds a-year, a member has the full benefit of the entire expenditure, with the range of a magnificent house, the use of a valuable library, reading-rooms, writing-rooms, billiard-rooms, smoking-rooms, baths, everything except beds. The propagation of Clubs has caused quite a revolution in the matter of town visitors' living. We saw that an unfortunate Boniface, who had got into the quagmire of the Insolvent Court, attributed his misfortunes to the altered system of the day, many of his once best customers, he said, now driving up to his door with their luggage, and after washing their hands adjourning to the Wellington, or the St. James's Hall instead of eating and drinking for the good of his house, as they used to do; but we know many men who have washed their hands of hotels altogether, and drive up to bachelor bedroom-houses in the neighbourhood of the Clubs, where for a few shillings a-night they get capitally lodged, with a sneck key and invisible valeting of the first order. Then having renovated their outer man on arriving, they go to their Clubs and live like princes, the best of everything being sought for their use.

Talk of country cream, country butter, country eggs, 'our farm of four acres', and so on; what country houses can surpass the butter, cream, and eggs of a first-rate London Club? Not only is the cream good, the butter good, and the eggs good, but the whole breakfast apparatus is of the nicest and most inviting order. Everything you want, and nothing more. Then the finely-flavoured tea is always so well made with real boiling water, instead of the lukewarm beverage we sometimes get; the muffins are fresh, the ham handsomely cut, the rolls crisp, and the toast neither leathery nor biscuity. A Club-breakfast is a meal to saunter over and enjoy, alternately sipping the tea and the newspaper.

The dinners are quite on a par with the breakfasts, and adapted to every variety of pocket and appetite. The best of all is, that though there is no previous arrangement on the part of the members, everything is as quickly supplied as if there had been. A quarter of an hour suffices to have dinner on the table—soup, fish, meats, sweets, and all.

Then the prices for which a man can live are something incredibly low; but as it is the nature of luxury to beget luxury, we do not know that the new generation have profited much, in a pecuniary point of view, by the establishment of Clubs. The old squires were rich—rich in the fewness of their wants, but the new squires have found wants that their forefathers were ignorant of. The old home manor won't do, they must have a moor; the row on the river won't do, they must have a yacht on the sea; the couple of hunters for Squire Jowleyman's hounds won't do, they must have six, and go upon grass; so that an increased expenditure has far more than absorbed the value of the reductions that have been made, and the money-saving advantages that have been acquired. The consequence of this is, that the new squires have begun to turn their attention to what their fathers had a great aversion to, namely, a little trade, and endeavour to 'make both ends meet', as Paul Pry used to say, by a little speculation. Railroads first led them astray at the time that all the world went mad together, and though it is true the Stock Exchange gentlemen were not so self-denying as to let any of the squires make any money at that time, yet the seed of the desire was sown, and has gone on fructifying ever since. Joint-Stock Banks were in favour until they brought so many parties down with a run, but the new Limited Liability Act offers great facilities for adventurous enterprise. We strongly suspect, however, that the squires will find no safer or better speculation than in draining and improving their own land. We do not advocate their teaching the farmers their trade, but we like to see them dispel the prejudices of habit by their example and superior intelligence.

Altogether the country gentlemen have become a very different race to what they were. They are more men of the world, and have shaken off the rancour and delusions of party, which, as Lord Brougham well said, 'allowed no merit in an adversary, and admitted no fault in a friend'. Whether this change is attributable to the emancipation of railways, or to the shock their system sustained by the ruthless repeal of the Corn-laws, or a combination of both, is immaterial to inquire.

(*Plain or Ringlets?* 1860.)

THE WOTHERSPOONS AWAIT THEIR GUESTS

BOTH Mr. and Mrs. Wotherspoon had risen sufficiently early to enable them to put the finishing stroke to their respective arrangements, and then to apparel themselves for the occasion. They were gorgeously attired, vieing with the rainbow in the colour of their clothes. Old Spoon, indeed, seemed as if he had put all the finery on he could raise, and his best brown cauliflower wig shone resplendent with Macassar oil. He had on a light brown coat with a rolling velvet collar, velvet facings and cuffs, with a magnificent green, blue, and yellow striped tartan velvet vest, enriched with red cornelian buttons, and crossed diagonally with a massive Brazilian gold chain, and the broad ribbon of his gold double-eye-glasses. He sported a light blue satin cravat, an elaborately worked ruby-studded shirt front, over a pink flannel vest, with stiff wrist-bands well turned up, showing the magnificence of his imitation India garnet buttons. On his clumsy fingers he wore a profusion of rings—a brilliant cluster, a gold and opal, a brilliant and sapphire, an emerald half-hoop ring, a massive mourning, and a signet ring—six in all—genuine or glass as the case might be, equally distributed between the dirty-nailed fingers of each hand. His legs were again encased in the treacherous white cords and woe-begone top-boots that were best under the breakfast table. He had drawn the thin cords on very carefully, hoping they would have the goodness to hang together for the rest of the day.

Mrs. Wotherspoon was bedizened with jewellery and machinery lace. She wore a rich violet-coloured velvet dress, with a beautiful machinery lace chemisette, fastened down the front with large Cairngorum buttons, the whole connected with a diminutive Venetian chain, which contrasted with the massive mosaic one that rolled and rattled upon her plump shoulders. A splendid imitation emerald and brilliant brooch adorned her bust, while her well-rounded arms were encircled with a mosaic gold, garnet and turquoise bracelet, an imitation rose diamond one, intermixed with pearl, a serpent armlet with blood-stone eyes, a heavy jet one, and an equally massive mosaic gold one with a heart's ease padlock. Though in the full development of womanhood, she yet distended her figure with crinoline, to the great contraction of her room.

(*Ask Mamma*, 1858.)

SOCIAL CALL

OLD Hall's house was in the heart of the town of Fleecyborough in Newbold-street, and, though substantial and well-built, could not vie with the more modern plateglass-windowed mansions that had sprung up in the outskirts and newer streets. It was a dingy brick mansion, with heavy woodwork windows, a massive green door, and an old iron railing enclosing nothing. Newbold-street at this part was rather narrow, and only flagged on Hall's side, but some fifty yards to the west was an airy market-place, and the bank, forming part of the house, was what was called extremely 'used' for business, the farmers popping in and out like rabbits in a warren. Though the bank was as dark and as dirty as a place could be, and the little partitioned-off nook, wherein we introduced the banker to our readers, was all the 'sweating room' he possessed, it was wonderful the amount of business he did, and the agonies parties underwent in that nook. 'Sivin and four's elivin, and nineteen's thirty—I'm afeard this bill won't do', Hall would say to a ponderous farmer who wanted a little accommodation, or perhaps a good deal, to enable him to meet his rent. 'Couldn't you get some 'un to join in a note?' or, to another, 'Sivin and four's elivin, and fifteen's twenty-six—it's not convenient just now', returning the gaping goose his hopeless paper. 'Ay—w-h-o-y—ar'll call again in haafe an hour', perhaps replies the innocent, not understanding the delicacy of the refusal.

But we are entering into the mysteries of Hall's calling, whereas our object is only to introduce his residence to our readers, preparatory to receiving company. We will now suppose our worthy friends in receipt of Lord Lavender's letter, and, the first transports of joy over, Mrs. Hall castle-building—imagining a match between our Tom and one of the Miss Myrtles, his lordship's daughters.

'Our Tom shall have an honourable for a wife!' exclaimed she.

'Sivin and four's elivin, and forty-one is fifty-two—I don't know that that would do him any good', replied Hall.

'Not do him any good!' retorted his wife; 'why, it's the very thing that Tom ought to have—a high-bred lady for a wife, who'll take him to court, and into distinguished society, and make a first-rate man of him.'

'Sivin and four's elivin, and eighty-three is ninety-four—I don't know that he'd be any better of that', replied the imperturbable banker.

'Not any better of *that*!' retorted his wife, who was all for advancement, and saw no reason why our Tummus should not marry a lord's daughter as well as Miss Nobody-knew-who Smith marry Lord Lavender; and so Hall and she got into a discussion on the point.

Their dialogue was interrupted by the most violent pounding of their hitherto peaceable, brass, lion-headed knocker, and before the astonished couple had recovered from the surprise, or speculated whether the bank was broke, or the house on fire, a second assault, if possible more furious than the first, thundered through the mansion, and caused a simultaneous rush to the drawing-room windows to see what was 'oop', as old Hall said. A tall, gold-laced hatted, moustachioed footman, in a dirty drab greatcoat, was in the act of returning to a high mail phaeton, yellow picked out with red, drawn by a pair of silver duns, in which was seated an enormous Daniel Lambert-looking man in undress uniform, and a little shrimp of a woman in a mixed costume of faded finery, in the shape of summer and winter clothes. A green terry-velvet bonnet with a yellow feather, a large ermine tippet over a light-blue muslin gown, with a machinery-lace-covered pink parasol, bright yellow-ochre-coloured gloves, and black velvet bands, with long ends and bright buckles round her wrists, as if she had sprained them. Altogether—man, woman, vehicle, horses—a very remarkable turn-out. The servant is now waiting for orders.

'ASK IF MISTRESS WHAT'S-HER-NAME'S AT HOME', bellowed the

monster, in a tone that sounded right into the house, and was heard by the curious on either side of the street, who had been attracted to their windows by the unwonted pounding of the door—'ASK IF MISTRESS WHAT'S-HER-NAME—HALL'S AT HOME', repeated he, catching the name, and flourishing his whip triumphantly over his stout Hanoverians.

'O *lauk*!' exclaimed Mrs. Hall in dismay. 'I'm not fit to be seen! I've got my old gown and a dirty cap on', glancing at herself in the eagle-topped mirror, as she hurried out of the room. 'Not at home, Sarey! not at home!' exclaimed she, leaning over the banisters to the maid, who, startled over the remains of a currant dumpling, was rushing pale and frightened to the door. 'Not at home, Sarey!—not at home', repeated Mrs. Hall, almost loud enough to be heard outside.

'Not at home!' blurted out Sarah, before the question was put at the half-opened door; and forthwith the lady in colours produced an elegant mother-o'-pearl card-case, and handed the footman an assortment of various sized cards for the not-at-homeites to help themselves to when they returned.

'Master's at home', observed Sarah, in a tremulous voice, with a laudable regard for the honour and credit of the bank.

'I THOUGHT YOU SAID NOT AT HOME', roared the officer, in a voice of thunder.

'Master *is*, missis is not', replied the maid timidly.

'AH—WELL, I'LL JUST GO IN AND SEE WHAT SORT OF A TIGER HE IS', observed the officer in the same tone after a pause; and, depositing the whip in its case, he handed the pipeclayed reins to the lady, and descended with a swag that shot her up in her seat like a pea.

He was indeed a fat man, and his crimson-and-gold belt was lost in the folds of fat at his sides. Having alighted on *terra firma*, he shook himself to see that he was all there, and then proceeded to labour in on his heels, paddling as it were with his short fat fins of arms.

The tiger had got himself into his lair ready for a pounce before the heavy man got creaked upstairs to the door which Sarey had left wide open, after a hurried half-frightened explanation of 'The gentleman, sir', hoping she was right in letting him in—fearing she was wrong.

'Sivin and four's elivin, and forty-five is fifty-six—what the deuce can the feller want with me?' muttered old Hall to himself. 'Sivin and four's elivin, and ninety-five's a 'under'd and six—he'll stand a dooced bad chance of gettin' a bill done after that impittence', thinking of his calling him a tiger. 'Sivin and four's elivin, and a 'under'd and fifteen is a 'under'd and twenty-six—what a time he is gettin' up', thought he, as the ponderous heavy-breathing man still laboured at the ascent. At length he appeared at the door.

'Mr. Hall (puff), I believe (wheeze)', gasped the officer, snatching his gold-laced foraging-cap off his great round head, and giving an uncouth bow with a kick out behind.

The banker acknowledged the impeachment without rising from his seat.

'I've called (puff)', roared he—'that's to say, Mrs. Colonel (wheeze) Blunt and (puff) I have done Mrs. (wheeze) Hall the (gasp) honour to call. I mean to say', continued he, waddling across the room to an easy-chair as he spoke—'I mean to say, Mrs. Colonel Blunt and (wheeze) I have done our (gasp) *selves* the honour to call on Mrs. (puff) What's-her-name', sousing himself into the chair as he spoke, 'to ask you to come to a little (puff) entertainment—music—mornin' hop, *thé dansant*, as she calls it, or ear-ache and stomach-ache, as I call it; and your (puff) son—how's your (puff) son? James, that's to say—fine young man (wheeze), great favourite of mine (puff); great (wheeze) pleasure in making his (gasp) 'quaintance. And your daughter; oh! I beg (puff) pardon, you haven't a daughter. It's Mr. Buss who has the daughter (puff); you townspeople are all so (puff) alike, you puzzle one. It's Mr. Buss who has the daughter—(puff)—dev'lish ugly girl she is too (wheeze);ugliest girl I ever saw—nasty-looking girl, I should say. He-he-he! Haw-haw-haw! Ho-ho-ho!'

Hall accompanied this speech, or rather parts of a speech, with the following mental commentary—

'Sivin and four's elivin, and forty-nine's sixty (what a fat man he is), and sixty's a 'under'd and twenty, and ninety's two 'under'd and ten (I wonder whether he'll be asking me to do a bill) and twenty-nine's two 'under'd and thirty-nine (that's a piece of impittence callin' Tummus, James—knows his name's Tummus as well as I do), and forty-five's two 'under'd and ninety-four (Miss Buss *is* an ugly girl)'; and as Hall hated old

Buss, the censure of the daughter rather expiated the offence of calling Tummus, James.

'Thank you, sir—that's to say, colonel—that's to say, sir—that's to say, Colonel Blunt', replied Hall, after the monster had exhausted himself. 'Mrs. H. and I are much obleged by the compliment of this call. *Tummus, not James*', continued Hall, eyeing the monster intently—'*Tummus, not James*', repeated he, 'will have much pleasure in accepting your note—that's to say, your *invitation*', continued he, with an emphasis, lest the inadvertency should lead to the production of a bill-stamp.

'Oh, but *you* must come too', roared the now recruited colonel; 'you must come too—*you* and *Mrs. What's-her-Name*, and all—hear my daughter play—finest performer in the world!—quite divine!'

'Sivin and four's elivin, and forty-eight's fifty-nine—there's a darter in the case, is there?' mused Hall. 'Thank you, sir—that's to say colonel', replied he, aloud. 'You're very good; but music's not much in my way.'

'Why, as to that', replied the colonel, with a shrug of his great shoulders—'why, as to that, I've no great eye for music myself; but the women like these sort of fandangoes, and we must knock under to them sometimes, you know—he, he, he!—haw, haw, haw!—ho, ho, ho!' his fat sides shaking like a shape of blancmange.

'Sivin and four's elivin, and eighty-three's ninety-four—my black shorts wouldn't show well by daylight', mused Hall, 'and Mrs. H. would be sure to want a new gown to go in. No, I thank you, Mister Colonel', resumed Hall, aloud; 'you're very good, but it's really quite out of my line, and Mrs. H., though very well at home, won't do to take abr*oo*ad.'

Just as Mr. Hall made this unfortunate declaration, the lady 'who didn't do to take abr*oo*ad' made her appearance, a splendidly revised edition of the one that had fled. A fine fly-away cap, with a full forty yards of pink ribbon, graced the back of her silvery-streaked head, while an elaborately-worked collar drooped over a shot-silk dress that assumed a variety of colours according to the light.

'Oh, here's Mrs. Buss!' exclaimed the colonel, as she entered; 'here's Mrs. Buss herself!'

'I say, Mrs. Buss, what d'ye think your husband says?'

roared the military monster, treating her just as he would a barmaid—'what d'ye think your husband says? He says, by Jove! that you're very well at home, but you don't do to take abroad—he, he, he! Now I should say', continued he, eyeing her intently—'I should say that you're a devilish deal better-looking woman than he is a man—haw, haw, haw!—ho, ho, ho! But, however, never mind', continued he, checking his guffaw; 'I'll tell you what I've come about—I'll tell you for what I've come about. Mistress Colonel Blunt and I have called to ask you to come to a *thé dansant,* or dancing tea, as she calls it; or ear-ache and stomach-ache, as I call it—you and your husband, and my friend Charles—so now you must come.'

He rose and rolled out of the room, leaving old Hall and his wife to settle the question of looks between them at their leisure as soon as they had recovered from the petrifaction of astonishment into which his condescending visit had thrown them. The Colonel then stumped downstairs, and climbing up into the phaeton, resumed the whip and reins, roaring out as he squashed himself into his seat, 'RUMMEST COUPLE I EVER SAW!' He then flourished the whip over the Hanoverians, the tall footman clambered up behind, and the rickety vehicle went jingling, like a tambourine, over the uneven pavement, to the delight of the children and the admiration of the country folks, who thought it a most splendid turn-out.

(*Young Tom Hall*, c.1851.)

CPSIA information can be obtained
at www.ICGtesting.com
Printed in the USA
BVHW031825111022
649159BV00004B/188